Electrodynamics Tutorials with Python Simulations

This book provides an accessible introduction to intermediate-level electrodynamics with computational approaches to complement a traditional mathematical treatment of the subject. It covers key topics in electrodynamics, such as electromagnetic fields, forces, potentials, and waves as well as Special Theory of Relativity.

Through intuition-building examples and visualizations in the Python programming language, it helps readers to develop technical computing skills in numerical and symbolic calculations, modeling and simulations, and visualizations. Python is a highly readable and practical programming language, making this book appropriate for students without extensive programming experience.

This book can serve as an electrodynamics textbook for undergraduate physics and engineering students in their second or third years, who are studying intermediate- or advanced-level electrodynamics and who want to learn techniques for scientific computing at the same time. This book will also appeal to computer science students who want to see how their computer programming skills may be applied to science, particularly to physics, without needing too much background physics knowledge.

Key features
- Major concepts in classical electrodynamics are introduced cohesively through computational and mathematical treatments.
- Computational examples in Python programming language guide students on how to simulate and visualize electrodynamic principles and phenomena for themselves.

Taejoon Kouh is a Professor of Nano and Electronic Physics at Kookmin University, Republic of Korea. He earned his B.A. in physics from Boston University and Sc.M. and Ph.D. degrees in physics from Brown University. After his study in Providence, RI, he returned to Boston, MA, and worked as a postdoctoral research associate in the Department of Aerospace and Mechanical Engineering at Boston University. He is a full faculty member in the Department of Nano and Electronic Physics at Kookmin University in Seoul, Korea, teaching and supervising undergraduate and graduate students. His current research involves the dynamics of nanoelectromechanical systems and the development of fast and reliable transduction methods and innovative applications based on tiny motion.

Minjoon Kouh is a program scientist for a philanthropic initiative. He was a Professor of Physics and Neuroscience at Drew University, USA, where he taught more than 30 distinct types of courses. He holds Ph.D. and B.S. degrees in physics from MIT and an M.A. from UC Berkeley. He completed a postdoctoral research fellowship at the Salk Institute for Biological Studies in La Jolla, CA. His research includes computational modeling of the primate visual cortex, information-theoretic analysis of neural responses, machine learning, and pedagogical innovations in undergraduate science education.

Series in Computational Physics

Series Editors: Steven A. Gottlieb and Rubin H. Landau

Parallel Science and Engineering Applications: The Charm++Approach
Laxmikant V. Kale, Abhinav Bhatele

Introduction to Numerical Programming: A Practical Guide for Scientists and Engineers Using Python and C/C++
Titus A. Beu

Computational Problems for Physics: With Guided Solutions Using Python
Rubin H. Landau, Manual José Páez

Introduction to Python for Science and Engineering
David J. Pine

Thermal Physics Tutorials with Python Simulations
Minjoon Kouh and Taejoon Kouh

Electrodynamics Tutorials with Python Simulations
Taejoon Kouh and Minjoon Kouh

For more information about this series, please visit: https://www.crcpress.com/Series-in-Computational-Physics/book-series/CRCSERCOMPHY

Electrodynamics Tutorials with Python Simulations

Taejoon Kouh and Minjoon Kouh

CRC Press
Taylor & Francis Group
Boca Raton London New York

CRC Press is an imprint of the
Taylor & Francis Group, an **informa** business

Designed cover image: Taejoon Kouh and Minjoon Kouh

First edition published 2024
by CRC Press
2385 NW Executive Center Drive, Suite 320, Boca Raton FL 33431

and by CRC Press
4 Park Square, Milton Park, Abingdon, Oxon, OX14 4RN

CRC Press is an imprint of Taylor & Francis Group, LLC

© 2024 Taejoon Kouh and Minjoon Kouh

Library of Congress Cataloging-in-Publication Data

Names: Kouh, Taejoon, author. | Kouh, Minjoon, author.
Title: Electrodynamics tutorials with Python simulations / Taejoon Kouh and Minjoon Kouh.
Description: First edition. | Boca Raton, FL : CRC Press, 2024. | Includes bibliographical references and index. | Summary: "This book provides an accessible introduction to intermediate-level electrodynamics with computational approaches to complement a traditional mathematical treatment of the subject. It covers key topics in electrodynamics, such as electromagnetic fields, forces, potentials, and waves as well as Special Theory of Relativity.
Identifiers: LCCN 2023048007 | ISBN 9781032498034 (hbk) | ISBN 9781032502311 (pbk) | ISBN 9781003397496 (ebk)
Subjects: LCSH: Electrodynamics--Textbooks. | Electrodynamics--Mathematical models--Textbooks. | Electrodynamics--Computer simulation--Textbooks. | Python (Computer program language)--Textbooks.
Classification: LCC QC631 .K68 2024 | DDC 537.60285/5133--dc23/eng/20231229
LC record available at https://lccn.loc.gov/2023048007

ISBN: 978-1-032-49803-4 (hbk)
ISBN: 978-1-032-50231-1 (pbk)
ISBN: 978-1-003-39749-6 (ebk)

DOI: 10.1201/9781003397496

Typeset in SFRM1000
by KnowledgeWorks Global Ltd.

Publisher's note: This book has been prepared from camera-ready copy provided by the authors.

To our family, Sarah, Chris, Cailyn, and Yumi

Contents

Preface

When we "do" physics, we employ a wide array of intellectual tools and methodologies. Mathematics stands as an essential and foundational tool, but the computational methods have become another powerful instrument. It is now important for physicists to equip themselves with the latest computational skills. We have written this tutorial to help readers learn basic skills in numerical and symbolic calculations, modeling and simulation, and data visualization using one of the most popular and versatile programming languages of the 2020s, Python, within the context of classical electrodynamics. If you are new to Python, the official homepage of the Python language (`www.python.org`) is a great place to visit.

This book covers key topics in electrodynamics at an undergraduate level, focusing on intuition-building examples and visualizations of electric and magnetic fields. After covering a few background topics on vectors and coding, we discuss the electric field in two dimensions for ease of visualization and comprehension and switch to the full three-dimensional formulation. Then, we introduce the magnetic field, followed by the topic of electromagnetic force and potential. Einstein's theory of special relativity is also included to show the relativity of electromagnetic forces. After a chapter on electromagnetic induction, we summarize the full set of Maxwell's equations in both integral and differential forms. The last chapter presents important consequences of Maxwell's equations in terms of the electromagnetic waves in free space. This book is focused mostly on the theory, but there are so many interesting examples around the applications of electrodynamics (e.g., motors, generators, transformers, electronic circuits, data storage, optics, etc.) that we encourage the readers to explore other resources on such topics. There are many excellent textbooks on electrodynamics, and some of our favorites are the ones written by Purcell, Griffiths, and Jackson.

About the authors

T. Kouh earned his BA in physics from Boston University and ScM and PhD in physics from Brown University. After his study in Providence, RI, he returned to Boston, MA, and worked as a postdoctoral research associate in the Department of Aerospace and Mechanical Engineering at Boston University. He is a full faculty member in the Department of Nano and Electronic Physics at Kookmin University in Seoul, Korea, teaching and supervising undergraduate and graduate students. His current research involves the dynamics of nanoelectromechanical systems and the development of fast and reliable transduction methods and innovative applications based on tiny motion.

M. Kouh's academic journey took place at MIT, UC Berkeley, Salk Institute, and Drew University. His research focused on computational modeling of the visual cortex, information-theoretic analysis of neural responses, machine learning, and pedagogical innovations in undergraduate science education. He taught more than 30 distinct types of courses at Drew as a professor in the physics and neuroscience programs. His other professional experiences include a role as a program scientist for a philanthropic initiative, a data scientist at a healthcare AI startup, and an IT consultant at a software company.

Acknowledgments

T. Kouh would like to thank his colleagues, both within the department and beyond, for their support, along with his students from the lab. Stimulating and engaging discussions with his students on various topics in physics have started him to mull over intriguing and entertaining ways of answering the questions, which has been valuable to completing this book. He is also grateful to his mentors for guiding him through his academic career and showing him the fun of doing physics. Last but not least, his biggest and deepest thanks go to his dearest Sarah.

M. Kouh would like to thank his mentors, colleagues, and students. They helped him to think deeply about physics from the first principles and from different perspectives. He is indebted to his academic mentors, who have shown the power and broad applicability of computational approaches, especially when they are thoughtfully combined with mathematical and experimental approaches. His family is a constant source of his inspiration and energy. Thank you, Yumi, Chris, and Cailyn!

## CHAPTER 1

Hills and Valleys

1.1 CREATING A RANGE OF VALUES WITH PYTHON

Most computational and numerical analyses of physical phenomena, including our study of electrodynamics, require working with a large set of numbers organized in various formats, such as tables, arrays, and vectors. In Python, we can access many useful data structures and mathematical operations using the `import` command. In the following code block, the line `import numpy as np` allows us to invoke useful, pre-programmed functions like `arange()`. For example, `x = np.arange(-5,7,2)` creates a one-dimensional array of values between −5 (inclusive) and 7 (exclusive) in steps of 2 and stores the array in a variable named x. Then, `print(x)` prints out the full array. We can address the individual elements in the array by using an index within square brackets `[]`. The first element of an array has an index of 0, so `x[0]` returns the first element in the array, −5, and `x[1]` returns the next element, −3. We can use negative integers to start indexing from the last elements. The last element has an index of −1, so `x[-1]` returns 5, and `x[-2]` returns 3.

```
# Code Block 1.1

import numpy as np

x = np.arange(-5,7,2)

print("Full array:",x)
print("First element:",x[0])
print("Second element:",x[1])
```

DOI: 10.1201/9781003397496-1

```
print("Last element:",x[-1])
print("Second to the last element:",x[-2])
```

```
Full array: [-5 -3 -1  1  3  5]
First element: -5
Second element: -3
Last element: 5
Second to the last element: 3
```

We can also address the range of elements by using :. For example, x[1:4] returns a sub-array with three elements: x[1], x[2], x[3], but not x[4]. If the starting index is omitted as in x[:4], the range is assumed to start from the very first element x[0]. If the ending index is omitted as in x[1:], the range is assumed to stretch to the very last element. We can also skip some indices by putting an additional :, so x[1:6:2] would return a sub-array with elements with indices 1, 3, and 5, in steps of 2, as shown in the following code block:

```
# Code Block 1.2

print(x[1:4]) # elements 1, 2, 3 (but not 4)
print(x[:4]) # elements 0, 1, 2, 3
print(x[1:]) # elements 1, 2, ..., -1.
print(x[:]) # Full array.
print(x[1:6:2]) # elements 1, 3, 5
```

```
[-3 -1  1]
[-5 -3 -1  1]
[-3 -1  1  3  5]
[-5 -3 -1  1  3  5]
[-3  1  5]
```

For plotting, we use various functions within the matplotlib module, as illustrated in the following code block. As usual, we import numpy and matplotlib modules and assign them the nicknames of np and plt, respectively. Such nicknames allow us to avoid potential name conflicts.

We define a range of data points between -1 and 1 in regular intervals with arange() function. By specifying an array ranging up to 1+step, we ensure that 1 is included as the last element in the array. The next line y = 2*(x**2)+1 specifies mathematical equality, $y = 2(x^2) + 1$.

The next few lines in the code block illustrate the usage of `scatter()` and `plot()` functions. The `scatter()` function puts markers at specified coordinates, and `plot()` draws straight lines between them. Hence, if the distance between the data points is small, the straight line segments produced by the `plot()` function will collectively make a smooth-looking curve.

The next lines in the code block improve the look of the plot. A good data visualization practice is to only include informative, relevant items and remove unnecessary, distracting ones in a plot. For example, three equally spaced tick-marks are enough to indicate the linear scale of the axis. Finer tick-marks or gridlines are unnecessary unless a precise reading of the coordinates is expected of a reader. Descriptive axis labels, titles, and legends also aid the readability of a plot. The goal is to increase the signal-to-noise ratio of our plots and visualizations.

```
# Code Block 1.3

import numpy as np
import matplotlib.pyplot as plt

# Define x and y.
step = 0.25
x = np.arange(-1,1+step,step)
y = 2*(x**2)+1

# Make a scatter plot with lines that connect adjacent points.
plt.scatter(x,y,color='gray')
plt.plot(x,y,color='black')

# Make the plot look nice by specifying limits and tick marks.
# Improve the signal-to-noise ratio of the visualization.
plt.xlim((-1.25,1.25))
plt.ylim((-1,5))
plt.xticks((-1,0,1))
plt.yticks(np.arange(0,5,2))
plt.xlabel('x')
plt.ylabel('y')

# Save the plot.
plt.savefig('fig_ch1_scatter_plot.eps')
plt.show()
```

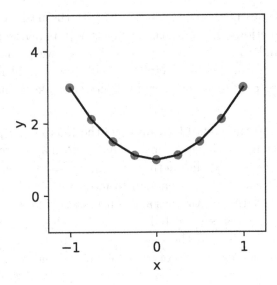

Figure 1.1

The `arange()` function in the numpy module is used to conveniently generate a one-dimensional array of equally spaced values. The `meshgrid()` function creates a two- or higher-dimensional array of values, as illustrated by the following code block.

```
# Code Block 1.4

step = 0.1
x = np.arange(-1,1+step,step)
y = np.arange(0,2+step,step)
xv, yv = np.meshgrid(x,y,indexing='ij')

plt.scatter(xv,yv,color='gray')
plt.xlim((-1.5,1.5))
plt.ylim((-0.5,2.5))
plt.xticks((-1,0,1))
plt.yticks((0,1,2))
plt.xlabel('x')
plt.ylabel('y')
plt.axis('equal')
plt.axis('square')

plt.savefig('fig_ch1_meshgrid.eps')
plt.show()
```

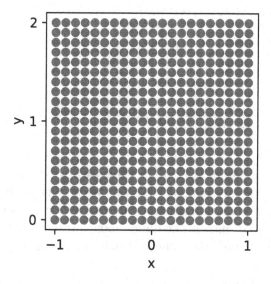

Figure 1.2

The optional argument `indexing` in the `meshgrid()` function specifies whether the returned array can be indexed and addressed like a matrix (row and then column) or like a Cartesian coordinate (horizontal and then vertical axes). The resulting arrays `xv` and `yv` are both two-dimensional matrices whose values collectively cover the full combination of ranges specified by `x` and `y`.

As a comparison, the following code block presents a more laborious and slower way of generating a two-dimensional array of numbers. We can start with an array of zeros whose shape is specified by the lengths of `x` and `y` arrays. This initialization is achieved by the `np.zeros()` function and the `len()` function that returns the length of each one-dimensional array. With two nested `for`-loops, we insert the values from the `x` and `y` arrays into the corresponding positions of the two-dimensional arrays. The resulting arrays, `x_slow` and `y_slow`, are identical to `xv` and `yv`. This identity can be checked with the `np.array_equal()` function. If two arrays are different, the `np.array_equal()` function would return false, and the `assert` command would generate an error message.

```
# Code Block 1.5

# A tedious way of generating a two-dimensional array.
# meshgrid() is more convenient.
```

```
x_slow = np.zeros((len(x),len(y)))
y_slow = np.zeros((len(x),len(y)))
for i in range(len(x)):
    for j in range(len(y)):
        x_slow[i,j] = x[i]
        y_slow[i,j] = y[j]

# Check if two different methods produce the same array.
assert np.array_equal(xv,x_slow)
assert np.array_equal(yv,y_slow)
```

Sometimes we may want to combine these two two-dimensional matrices (xv and yv) into a list of pairs p, allowing us to refer to all x coordinates simply as p[0] and all y coordinates as p[1]. We can reshape each two-dimensional matrix into a one-dimensional array with the `flatten()` method and then stack them together with the `vstack()` function.

```
# Code Block 1.6
p = np.vstack((xv.flatten(),yv.flatten()))
# With the array p, the same plot can be created with this command:
# plt.scatter(p[0],p[1],color='gray')
```

1.2 LANDSCAPE OF GAUSSIANS

Imagine a landscape of hills and valleys. We can use a Gaussian function as a mathematical model for a hill and an inverted Gaussian function for a valley. A Gaussian is defined as

$$f(x) = e^{-\frac{(x-x_0)^2}{2\sigma^2}},$$

where x_0 determines the location of the peak and σ determines its width.

The following code block defines and plots a one-dimensional Gaussian function. A hill f is located at $x_0 = +0.3$ with $\sigma = 0.2$. A valley g is at $x_0 = -0.3$ with $\sigma = 0.3$.

```
# Code Block 1.7

import matplotlib.pyplot as plt
import numpy as np

step = 0.01
x = np.arange(-1,1,step)

f = +np.exp(-(x-0.3)**2/2/0.2**2)
g = -np.exp(-(x+0.3)**2/2/0.3**2)
f_label = r'$+e^{-\frac{(x-0.3)^2}{2 \times 0.2^2}}$'
g_label = r'$-e^{-\frac{(x+0.3)^2}{2 \times 0.3^2}}$'
```

```
fig = plt.figure(figsize=(5,3))
plt.plot(x,f,label=f_label,color='black')
plt.plot(x,g,label=g_label,color='gray')
plt.legend(framealpha=1)
plt.xticks((-1,0,1))
plt.yticks((-1,0,1))
plt.ylim((-1.2,1.2))
plt.xlabel('x')
plt.ylabel('y')
plt.savefig('fig_ch1_gauss_1d.eps')
plt.show()

# Make a plot of a sum of two Gaussian functions.

h = f+0.5*g
h_label = r'$e^{-\frac{(x-0.3)^2}{2 \times 0.2^2}}$'
h_label = h_label + r'$-0.5e^{-\frac{(x+0.3)^2}{2 \times 0.3^2}}$'

fig = plt.figure(figsize=(5,3))
plt.plot(x,h,color='black',label=h_label)
plt.legend(framealpha=1)
plt.xticks((-1,0,1))
plt.yticks((-1,0,1))
plt.ylim((-1.2,1.2))
plt.xlabel('x')
plt.ylabel('y')
plt.savefig('fig_ch1_gauss_1d_sum.eps')
plt.show()
```

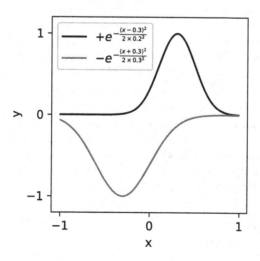

Figure 1.3

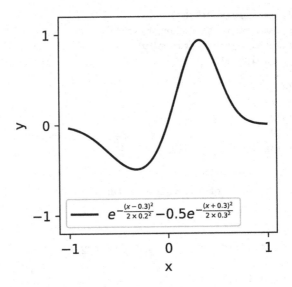

Figure 1.4

A two-dimensional Gaussian function with a peak located at (x_o, y_o) is defined as

$$g(x, y) = e^{-\frac{(x-x_o)^2+(y-y_o)^2}{2\sigma^2}}$$

and can be plotted as shown in the following code block:

```python
# Code Block 1.8

step = 0.05
x, y = np.meshgrid(np.arange(-1,1,step),
                   np.arange(-1,1,step),
                   indexing='ij')
x0, y0 = 0.3, 0.1
sig = 0.3
g = np.exp(-((x-x0)**2+(y-y0)**2)/2/sig**2)

fig = plt.figure(figsize=(5,4))
ax = plt.axes(projection='3d')
ax.plot_surface(x,y,g,cmap='gray')
ax.set_xlabel('x')
ax.set_ylabel('y')
ax.set_zlabel('z')
ax.set_xticks((-1,0,1))
ax.set_yticks((-1,0,1))
ax.set_zticks((0,0.5,1))
ax.view_init(10, -60)
```

```
ax.set_rasterized(True)
plt.tight_layout()
plt.savefig('fig_ch1_gauss_2d.eps')
plt.show()
```

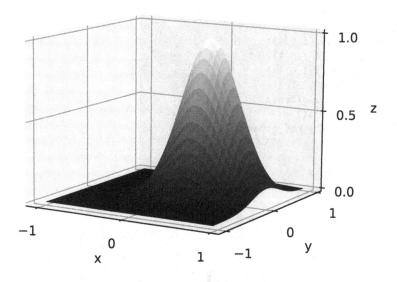

Figure 1.5

We can add more Gaussians to make the resulting landscape a bit more interesting.

```
# Code Block 1.9

# Make a 3D plot of a landscape.

def sample_sum_of_gauss(step=0.01):
    x, y = np.meshgrid(np.arange(-1,1,step),
                       np.arange(-1,1,step),
                       indexing='ij')
    x0,y0,x1,y1,x2,y2 = -0.3,-0.2,0.3,0.1,0.0,0.1
    sig0,sig1,sig2 = 0.3,0.2,0.5
    g0 = +3.0*np.exp(-((x-x0)**2+(y-y0)**2)/2/sig0**2)
```

```
    g1 = +2.0*np.exp(-((x-x1)**2+(y-y1)**2)/2/sig1**2)
    g2 = -2.0*np.exp(-((x-x2)**2+(y-y2)**2)/2/sig2**2)
    z = g0+g1+g2
    return x,y,z

x,y,z = sample_sum_of_gauss()
fig = plt.figure(figsize=(5,4))
ax = plt.axes(projection='3d')
ax.plot_surface(x,y,z,cmap='gray')
ax.set_xlabel('x')
ax.set_ylabel('y')
ax.set_zlabel('z')
ax.set_xticks((-1,0,1))
ax.set_yticks((-1,0,1))
ax.set_zticks((-1,0,1))
ax.view_init(30, -70)
ax.set_rasterized(True)
plt.tight_layout()
plt.savefig('fig_ch1_gauss_2d_sum.eps')
```

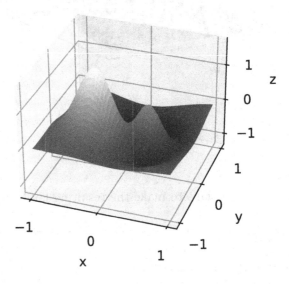

Figure 1.6

We can look at the landscape with a bird's eye view (from the positive z-axis) by using the plotting command `plt.contourf()`.

```
# Code Block 1.10

# Make a 2D contour plot.

fig = plt.figure(figsize=(4,4))
plt.contourf(x,y,z,cmap='gray')
plt.xlim((-2,2))
plt.ylim((-2,2))
plt.axis('square')
plt.axis('equal')
plt.axis('off')
plt.savefig('fig_ch1_gauss_2d_sum_contour.eps')
plt.show()
```

Figure 1.7

Now imagine steady rain falling onto this landscape of hills and valleys. When the raindrops hit the ground, they start flowing downhill, guided by the ups and downs of the landscape. As water flows from higher to lower grounds, it moves quickly down a steep slope and slowly down a gentle slope. The velocity of water at a different position can be represented by an arrow that points in the direction of the downward slope. The length of each arrow can represent the steepness of the landscape at that position.

The following code block visualizes this idea using the functions `np.gradient()` and `plt.quiver()`. We will have much more to say about these functions in later chapters. For now, just focus on the plot, not the code for generating it. The contour plot shows the highs and lows of the landscape, and the overlaid arrows indicate the steepness and the direction of the slopes at different positions in the region.

```python
# Code Block 1.11

# Show the gradient with quiver.
fig = plt.figure(figsize=(4,4))
plt.contour(x,y,z,cmap='gray')

# Coarse version of x,y,z
xc,yc,zc = sample_sum_of_gauss(step=0.1)
u,v = np.gradient(zc)
plt.quiver(xc,yc,-u,-v)
plt.xlim((-2,2))
plt.ylim((-2,2))
plt.axis('square')
plt.axis('equal')
plt.axis('off')
plt.savefig('fig_ch1_gauss_2d_sum_grad.eps')
plt.show()
```

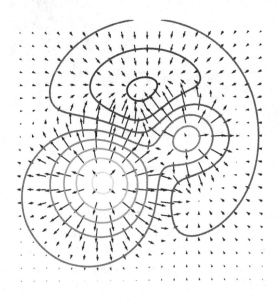

Figure 1.8

Here is a different analogy. The above plot may be interpreted as a weather map of a region. The contours represent the air pressure, and the arrows represent the winds. Some subregions may have higher pressure, and others may have lower pressure. Wind will blow from high- to low-pressure regions. A bigger pressure difference will cause a stronger wind. If you launch air balloons at different positions, they will start moving along the arrows.

In our study of electrodynamics, we will take advantage of these intuitions, as they are particularly useful for developing the concepts of electromagnetic potentials and fields.

## CHAPTER 2

Vector

Whether it is a raindrop rolling down a hill or wind blowing due to the difference in air pressure, we often need to work with both magnitude (fast versus slow or strong versus weak) and direction (from high to low or from north to south) of a vector quantity. The following is a non-comprehensive review of vectors and their mathematical operations. For a more comprehensive introduction, we refer the readers to a textbook on linear algebra and vector calculus.

A vector in three dimensions can be written as:

$$\vec{A} = a_x\hat{x} + a_y\hat{y} + a_z\hat{z} = \left(a_x, a_y, a_z\right),$$

where a_x is the magnitude of the vector's component along the direction of x-axis. Similarly, a_y and a_z are the components along y and z-axes. The magnitude of a vector is defined as:

$$|\vec{A}| = \sqrt{a_x^2 + a_y^2 + a_z^2}.$$

A unit vector is a vector of magnitude 1, and it is denoted with a hat symbol. Hence, the three orthogonal directions along the x, y, and z axes are represented as $\hat{x}$, $\hat{y}$, and $\hat{z}$, respectively. We can make any vector into a unit vector by dividing it by its magnitude.

$$\hat{A} = \frac{\vec{A}}{|\vec{A}|} = \frac{a_x}{\sqrt{a_x^2 + a_y^2 + a_z^2}}\hat{x} + \frac{a_y}{\sqrt{a_x^2 + a_y^2 + a_z^2}}\hat{y} + \frac{a_z}{\sqrt{a_x^2 + a_y^2 + a_z^2}}\hat{z}.$$

DOI: 10.1201/9781003397496-2

Two vectors can be added by summing up their components along each dimension, and the result will also be a vector.

$$\vec{A} + \vec{B} = (a_x + b_x)\hat{x} + (a_y + b_y)\hat{y} + (a_z + b_z)\hat{z}.$$

Subtraction is the addition of an inverse, or component-wise subtractions.

$$\vec{A} - \vec{B} = \vec{A} + (-\vec{B}) = (a_x - b_x)\hat{x} + (a_y - b_y)\hat{y} + (a_z - b_z)\hat{z}.$$

2.1 DRAWING A VECTOR WITH PYTHON

The `plt.quiver()` function allows us to draw an arrow as a representation of a vector. The first two input arguments to this function designate the position of the tail of a vector, and the next two input arguments are the corresponding vector components in the x and y directions. As illustrated in the following code block, the command `plt.quiver(0,0,A[0],A[1])` draws a vector whose tail is located at the origin and whose components are given by `A[0]` and `A[1]`. In this example, $\vec{A} = 1\hat{x} + 2\hat{y}$.

```
# Code Block 2.1

import numpy as np
import matplotlib.pyplot as plt

# Let us draw a vector with the quiver() function.
# Note: the extra arguments (angles, scale, ...)
# ensures that the arrow lengths are properly scaled.
A = np.array([1,2])
plt.quiver(0,0,A[0],A[1],angles='xy',scale_units='xy',scale=1)
plt.text(A[0]+0.1,A[1]+0.1,r"$\vec{A}$")

B = np.array([1,-2])
plt.quiver(0,0,B[0],B[1],angles='xy',scale_units='xy',scale=1)
plt.text(B[0]+0.1,B[1]+0.1,r"$\vec{B}$")

C = np.array([-2,-1])
plt.quiver(0,0,C[0],C[1],angles='xy',scale_units='xy',scale=1)
plt.text(C[0]+0.1,C[1]-0.5,r"$\vec{C}$")

plt.grid()
plt.axis('square')
plt.xlabel('x')
```

```
plt.ylabel('y')
lim = 3
plt.xlim((-lim,lim))
plt.ylim((-lim,lim))
plt.xticks(np.arange(-lim,lim+0.1))
plt.yticks(np.arange(-lim,lim+0.1))
plt.savefig('fig_ch2_vector_quiver.eps')
plt.show()
```

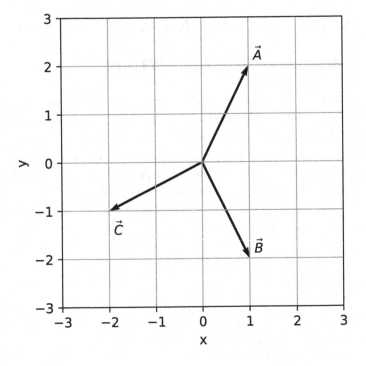

Figure 2.1

2.2 VECTOR PRODUCTS

A dot product, also known as a scalar or inner product, between two vectors $\vec{\mathbf{A}}$ and $\vec{\mathbf{B}}$ is defined as

$$\vec{\mathbf{A}} \cdot \vec{\mathbf{B}} = a_x b_x + a_y b_y + a_z b_z = \sum_{i=1}^{d} a_i b_i,$$

where in the last expression, each dimensional component is identified by an integer index i and the summation goes through all d dimensions.

As the name suggests, the result of this operation is a scalar value, not a vector. It can also be proven that $\vec{A} \cdot \vec{B} = |\vec{A}||\vec{B}| \cos \theta$, where θ is the angle between the two vectors.

A cross product, also known as vector product, between vector $\vec{A}$ and $\vec{B}$ is defined in three dimensions as:

$$\vec{A} \times \vec{B} = \begin{vmatrix} \hat{x} & \hat{y} & \hat{z} \\ a_x & a_y & a_z \\ b_x & b_y & b_z \end{vmatrix} = (a_y b_z - a_z b_y)\hat{x} + (a_z b_x - a_x b_z)\hat{y} + (a_x b_y - a_y b_x)\hat{z},$$

which is a vector quantity. It can be proven that $\vec{A} \times \vec{B} = |\vec{A}||\vec{B}| \sin \theta \, \hat{n}$, where θ is the angle between the two vectors and $\hat{n}$ is a unit vector that is orthogonal to the plane made by $\vec{A}$ and $\vec{B}$.

The specific direction of $\hat{n}$ can be found with your right hand. As you curl your four right fingers from $\vec{A}$ to $\vec{B}$, $\hat{n}$ will be in the direction of your right thumb. The cross-product is anti-commutative, meaning that $\vec{A} \times \vec{B} = -\vec{B} \times \vec{A}$, so the order of this vector operation matters. The np.cross() function calculates a cross product.

```
# Code Block 2.2

# Visualizing the cross product

A = np.array([2,1,0])
B = np.array([0,0,1])
C = np.cross(A,B)

x, z = np.meshgrid(np.linspace(-1,1,5), np.linspace(-1,1,5))
y = x/2 # Define a plane formed by A and B.

fig = plt.figure(figsize=(5,5))
ax = plt.axes(projection='3d')
ax.plot_surface(x,y,z,color='#CCCCCC',alpha=0.2)
ax.quiver(0,0,0,A[0],A[1],A[2],color='k',
          arrow_length_ratio=0.1,normalize=True)
ax.quiver(0,0,0,B[0],B[1],B[2],color='k',
          arrow_length_ratio=0.1,normalize=True)
ax.quiver(0,0,0,C[0],C[1],C[2],color='k',
          arrow_length_ratio=0.1,normalize=True)
ax.text(1,0.5,0,r"$\vec{A}$")
ax.text(0,0,1.1,r"$\vec{B}$")
ax.text(0.5,-1,0,r"$\vec{A}\times\vec{B}$")
ax.set_xticks((-1,0,1))
```

```
ax.set_yticks((-1,0,1))
ax.set_zticks((-1,0,1))
ax.set_xlim(-1, 1)
ax.set_ylim(-1, 1)
ax.set_zlim(-1, 1)
ax.set_xlabel('x')
ax.set_ylabel('y')
ax.set_zlabel('z')
ax.view_init(20,-120)
plt.savefig('fig_ch2_vector_cross.eps')
```

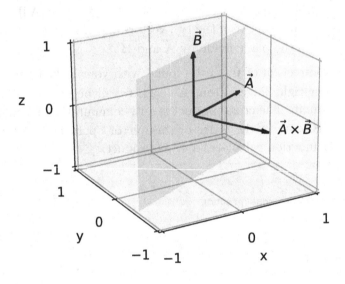

Figure 2.2

There is no division operation between two vectors because there isn't a mathematically meaningful way to divide one direction by another.

2.3 VECTOR DECOMPOSITION

Finding the individual components of a vector is called vector decomposition. The vector components along the x and y axes of a two-dimensional vector, $\vec{\mathbf{A}}$, can be calculated with the **cos** and **sin** functions, based on the magnitude (r) and directional angle (ϕ) of the vector, where ϕ is conveniently chosen as the counterclockwise angle from the positive

x-axis. Then, $a_x = r \cos \phi$ and $a_y = r \sin \phi$, as illustrated by the following code block.

Using a dot product operation, $a_x = \vec{\mathbf{A}} \cdot \hat{\mathbf{x}}$ and $a_y = \vec{\mathbf{A}} \cdot \hat{\mathbf{y}}$. In other words, the vector component along each coordinate axis can be determined by projecting the vector onto each axis and measuring the extent of this projection.

```
# Code Block 2.3

# Vector decomposition

# defining the famous constant pi = 3.14...
pi = np.pi

r = 1
phi_circle = np.arange(0,2*pi,0.01)
x_circle = r*np.cos(phi_circle)
y_circle = r*np.sin(phi_circle)

phi = pi/6
x = r*np.cos(phi)
y = r*np.sin(phi)

plt.figure(figsize=(5,5))
plt.plot(x_circle,y_circle,color='gray')
plt.quiver(0,0,x,y,angles='xy',scale_units='xy',scale=1)
plt.quiver(0,0,x,0,angles='xy',scale_units='xy',scale=1)
plt.quiver(0,0,0,y,angles='xy',scale_units='xy',scale=1)

plt.plot([0,0],[-r,r],linestyle='dotted',color='gray')
plt.plot([-r,r],[0,0],linestyle='dotted',color='gray')

plt.plot([0,x],[y,y],linestyle='dotted',color='gray')
plt.plot([x,x],[0,y],linestyle='dotted',color='gray')
plt.text(0.3,-0.15,r"$a_x = r\ \cos \phi$")
plt.text(-0.1,0.6,r"$a_y = r\ \sin \phi$")
plt.text(0.4,0.3,"r")
plt.text(0.25,0.05,r"$\phi$")
plt.text(1.1,0,"x")
plt.text(0,1.1,"y")
plt.axis('square')
plt.axis('off')
plt.xlim(np.array([-1,1])*r*1.1)
plt.ylim(np.array([-1,1])*r*1.1)
plt.savefig('fig_ch2_vector_decompose.eps')
plt.show()
```

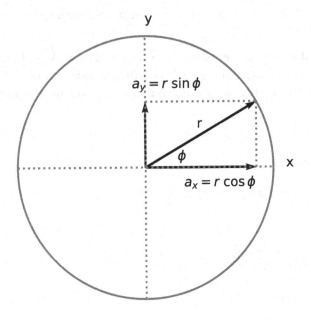

Figure 2.3

2.4 VECTOR CALCULUS

Calculus provides a mathematically precise and concise way to describe motion. In classical mechanics, the motion of an object undergoing constant acceleration a, like a free-falling mass under gravity, can be described with a set of kinematics equations:

$$
\begin{aligned}
x(t) &= x_o + v_o t + \frac{1}{2}at^2 \\
v(t) &= v_o + at,
\end{aligned}
$$

where the object's velocity changes linearly in time, and its position changes quadratically in time.

The relationship between velocity $v(t)$ and position $x(t)$ is that of a derivative and antiderivative, so if you know one quantity, the other can be determined.

$$\text{If you know } x(t), \quad v(t) = \frac{dx}{dt}.$$

$$\text{If you know } v(t) \text{ and } x_0, \quad x(t) = \int_0^t v(t')dt' + x_0.$$

In other words, the movement of an object can be formulated in terms of a derivative or an integral of some physical quantities. Either formulation, whether it starts with $x(t)$ or $v(t)$, describes the same motion. Similarly, the mathematical descriptions of electromagnetism can be formulated in terms of either integrals or derivatives, as we will see in later chapters. In this section, we present the definitions and examples of a few essential vector calculus operations.

Given a three-dimensional scalar function $f(x, y, z)$, its spatial rate of change along the x-axis is given by $\frac{\partial f}{\partial x}$. Given a three-dimensional vector function $\vec{K} = K_x \hat{x} + K_y \hat{y} + K_z \hat{z}$, we may consider the sum of the rate of change along each dimension, by calculating $\frac{\partial K_x}{\partial x} + \frac{\partial K_y}{\partial y} + \frac{\partial K_z}{\partial z}$. The physical interpretation of those derivatives will be discussed further when we encounter them again in the context of electric and magnetic fields. For now, we introduce a vector operator called del or nabla, ∇ for expressing the derivatives in higher dimensions more compactly and elegantly. This operator is defined as $\nabla = \frac{\partial}{\partial x}\hat{x} + \frac{\partial}{\partial y}\hat{y} + \frac{\partial}{\partial z}\hat{z}$. If this operator is applied to a scalar function $f(x, y, z)$, we obtain a new vector quantity called the gradient:

$$\text{Gradient: } \nabla f(x, y, z) = \frac{\partial f}{\partial x}\hat{x} + \frac{\partial f}{\partial y}\hat{y} + \frac{\partial f}{\partial z}\hat{z}.$$

This operator can be applied to a vector function, too. Given a vector function $\vec{K}$, its divergence, which is a scalar quantity, is defined as

$$\text{Divergence: } \nabla \cdot \vec{K} = \frac{\partial K_x}{\partial x} + \frac{\partial K_y}{\partial y} + \frac{\partial K_z}{\partial z}.$$

The curl of $\vec{K}$ is a vector quantity, and it is defined as

$$\text{Curl: } \nabla \times \vec{K} = \begin{vmatrix} \hat{x} & \hat{y} & \hat{z} \\ \frac{\partial}{\partial x} & \frac{\partial}{\partial y} & \frac{\partial}{\partial z} \\ K_x & K_y & K_z \end{vmatrix} = \left(\frac{\partial K_z}{\partial y} - \frac{\partial K_y}{\partial z} \right)\hat{x} + \left(\frac{\partial K_x}{\partial z} - \frac{\partial K_z}{\partial x} \right)\hat{y} + \left(\frac{\partial K_y}{\partial x} - \frac{\partial K_x}{\partial y} \right)\hat{z}.$$

Doing integrals in higher dimensions takes additional care. For example, given a three-dimensional scalar function $f(x, y, z)$, we may perform a volume integral within a particular region enclosed by a two-dimensional boundary. The notation for such an integral is

$$\oiiint_{\text{Volume}} f(x, y, z) dx dy dz,$$

where the circle around the integral symbol signifies an integral over an enclosed volume. Similarly, we may sometimes deal with an integral over a two-dimensional surface or an integral along a one-dimensional contour. Such integrals will be introduced and explained with relevant contexts in the upcoming chapters.

The last topic in this section is the fundamental theorems of vector calculus. According to the more-familiar fundamental theorem of calculus,

$$\int_a^b \frac{df(x)}{dx} dx = f(b) - f(a),$$

which states that the integral of a derivative over a range can be calculated with the values of the antiderivative at the range's boundary (i.e., the endpoints a and b). It elegantly captures the relationships between a function, its derivative, integral, and boundary condition.

Similarly, the fundamental theorem of gradient states that:

$$\int_{\vec{a}}^{\vec{b}} \nabla f \cdot d\vec{l} = f(\vec{b}) - f(\vec{a}).$$

As an intuition-building example, consider a conservative force $\vec{F}$ like gravity. Imagine performing work under this force so that we are moving an object from point $\vec{a}$ to point $\vec{b}$. The amount of work is given by $\int_{\vec{a}}^{\vec{b}} \vec{F} \cdot d\vec{l}$, which equals the change of the object's potential energy $\Delta U = U(\vec{b}) - U(\vec{a})$. In the absence of friction (i.e., with the assumption of conservative force), the amount of work only depends on the final and initial points of the object, and it does not matter how the object ended up at its final position. In other words, the change in potential energy is the same whether the object was brought from $\vec{a}$ to $\vec{b}$ in a straight or curved path. In the case of an object with mass m displaced by height h near the earth's surface, this is summarized as a formula for the gravitational potential energy, $\Delta U = mgh$.

The divergence and curl have their own fundamental theorems. The fundamental theorem of divergence states that the volume integral of the divergence of a vector function equals the surface integral of the vector function over the surface enclosing the volume of interest. For the curl of a vector function, its surface integral equals the line integral of the vector function over the closed linear contour defining this surface. The later chapters will go over their meanings and applications. For now, we present them below in one place for easy reference and comparison.

$$\iiint_{\text{Volume}} (\nabla \cdot \vec{K}) dV = \oiint_{\text{Area}} \vec{K} \cdot d\vec{a}.$$

$$\iint_{\text{Area}} (\nabla \times \vec{K}) \cdot d\vec{a} = \oint_{\text{Contour}} \vec{K} \cdot d\vec{l}.$$

2.5 ADDITIONAL OPERATIONS WITH ∇

Just as we have a rule for differentiating a product of two functions, $\frac{d}{dt}(f(t)g(t)) = f(t)\frac{dg(t)}{dt} + g(t)\frac{df(t)}{dt}$, there are product rules for the ∇-operator, when it is applied to the scalar and vector functions. For instance, if you have a dot product of two vector functions $\vec{K}$ and $\vec{G}$, its gradient will be given by:

$$\nabla(\vec{K} \cdot \vec{G}) = \vec{K} \times (\nabla \times \vec{G}) + \vec{G} \times (\nabla \times \vec{K}) + (\vec{K} \cdot \nabla)\vec{G} + (\vec{G} \cdot \nabla)\vec{K}.$$

Given a cross product between two vector functions, its divergence and curl can be determined by

$$\nabla \cdot (\vec{K} \times \vec{G}) = \vec{G} \cdot (\nabla \times \vec{K}) - \vec{K} \cdot (\nabla \times \vec{G})$$

and

$$\nabla \times (\vec{K} \times \vec{G}) = (\vec{G} \cdot \nabla)\vec{K} - (\vec{K} \cdot \nabla)\vec{G} + \vec{K}(\nabla \cdot \vec{G}) - \vec{G}(\nabla \cdot \vec{K}).$$

These product rules are helpful when evaluating complex vector relations.

Second derivatives are also possible with the ∇-operator. The divergence of the gradient of a scalar function is given by:

$$\nabla \cdot \nabla f(x, y, z) = \nabla^2 f(x, y, z) = \frac{\partial^2 f}{\partial x^2} + \frac{\partial^2 f}{\partial y^2} + \frac{\partial^2 f}{\partial z^2}.$$

The curl of the curl of a vector function is also a sensible quantity. It can be proved that

$$\nabla \times (\nabla \times \vec{K}) = \nabla(\nabla \cdot \vec{K}) - \nabla^2 \vec{K}.$$

We note that not all combinations of ∇ operations make sense. For example, $\nabla \times (\nabla \cdot \vec{K})$ is not a valid mathematical expression, since the divergence $\nabla \cdot \vec{K}$ is not a vector function, so we cannot calculate its curl. Similarly, $\nabla(\nabla \times \vec{K})$, or the gradient of the curl of a vector function, is not a valid expression since the curl of a vector function is a vector function and gradient is not defined on a vector function.

## CHAPTER 3

Vector Field

The map of the arrows we have seen at the end of the first chapter
is called a vector field, where an arrow (a vector quantity with both
a direction and a magnitude) is specified at every position. The term
vector field is often interchangeable with the term vector function, but
it emphasizes that the domain of the function is spatial. These vectors
do not change over time on a static vector field. On a dynamic vector
field, these vectors can change and evolve over time. A dynamic vector
field is like a changing landscape or a wind map under changing weather.
We will first discuss static vector fields over the next few chapters and
then work with dynamic vector fields later.

3.1 SIMPLE VECTOR FIELDS

The following code block creates a simple vector field where at each
point in space, the vector is pointing in the positive $\hat{x}$ direction. Since
the vectors are identical everywhere, this vector field is uniform (i.e., no
spatial dependence) and static (i.e., no temporal dependence).

```
# Code Block 3.1
import numpy as np
import matplotlib.pyplot as plt

step = 0.25

# Set up a grid of (x,y) coordinates
x,y = np.meshgrid(np.arange(-1,1+step,step),
                  np.arange(-1,1+step,step),
                  indexing='ij')
```

DOI: 10.1201/9781003397496-3

```
dx = 1
dy = 0
fig = plt.figure(figsize=(2,2))
plt.quiver(x,y,dx,dy,angles='xy',scale_units='xy')
plt.axis('square')
plt.axis('off')
plt.xlim(np.array([-1,1])*1.1)
plt.ylim(np.array([-1,1])*1.1)
plt.savefig('fig_ch3_simple_field.pdf')
plt.show()
```

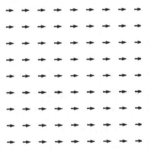

Figure 3.1

We can create other examples of uniform and static vector fields, as shown below.

```
# Code Block 3.2

f, axes = plt.subplots(2,2,figsize=(6,6))

ax = axes[0,0]
ax.set_title('(a) left')
ax.quiver(x,y,-1,0,angles='xy',scale_units='xy')

ax = axes[0,1]
ax.set_title('(b) up')
ax.quiver(x,y,0,1,angles='xy',scale_units='xy')

ax = axes[1,0]
ax.set_title('(c) up-right')
ax.quiver(x,y,1,1,angles='xy',scale_units='xy')

ax = axes[1,1]
ax.set_title('(d) down-right')
```

```
ax.quiver(x,y,1,-1,angles='xy',scale_units='xy')

for i in range(2):
    for j in range(2):
        ax = axes[i,j]
        ax.axis('square')
        ax.axis('off')
        ax.set_xlim(np.array([-1,1])*1.1)
        ax.set_ylim(np.array([-1,1])*1.1)

plt.tight_layout()
plt.savefig('fig_ch3_other_fields_1.pdf')
plt.show()
```

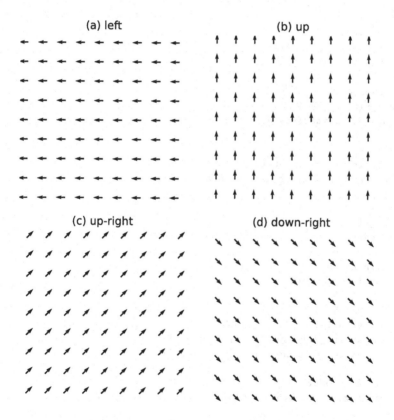

Figure 3.2

Other patterns of vector fields are possible. The following code block shows a few examples of non-uniform vector fields where (a) the vectors

are pointing radially away from the origin with equal magnitudes and
(b) the vectors are pointing radially, but their magnitudes decrease as $\frac{1}{r}$
with $r = \sqrt{x^2 + y^2}$. In (c) and (d), the vectors are swirling in clockwise
and counterclockwise directions.

```python
# Code Block 3.3

# Add a small number to avoid dividing by zero.
small_number = 10**(-10)
r = np.sqrt(x**2+y**2)+small_number
xhat = x/r
yhat = y/r

f, axes = plt.subplots(2,2,figsize=(6,6))

ax = axes[0,0]
ax.set_title('(a) radial')
ax.quiver(x,y,xhat,yhat,angles='xy',scale_units='xy')

ax = axes[0,1]
ax.set_title('(b) 1/r')
ax.quiver(x,y,xhat/r,yhat/r,angles='xy',scale_units='xy')

ax = axes[1,0]
ax.set_title('(c) cw swirl')
ax.quiver(x,y,y,-x,angles='xy',scale_units='xy')

ax = axes[1,1]
ax.set_title('(d) ccw swirl')
ax.quiver(x,y,-y,x,angles='xy',scale_units='xy')

for i in range(2):
    for j in range(2):
        ax = axes[i,j]
        ax.axis('square')
        ax.axis('off')
        ax.set_xlim(np.array([-1,1])*1.1)
        ax.set_ylim(np.array([-1,1])*1.1)

plt.tight_layout()
plt.savefig('fig_ch3_other_fields_2.pdf')
plt.show()
```

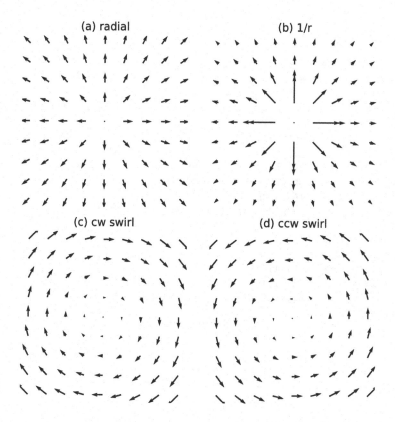

Figure 3.3

3.2 FLUX

We might ask questions like "Can we numerically characterize the overall pattern of a vector field?" or "How much flow is there in a vector field?" Let us start with the question about the amount of flow. As shown in the following figure, consider a vertical line placed inside a uniform vector field pointing to the right. A few sample vectors are drawn with their starting points on the line. This line is not a physical object, so it does not block or disturb the flow of the vector field. It simply marks a region of analysis, allowing us to talk about the amount of the vector field's flow. We will also refer to it as a boundary. A boundary may be straight, curved, or closed to form a shape like a circle or a square. When dealing with a three-dimensional vector field, such a boundary would be a two-dimensional surface.

```
# Code Block 3.4
fig = plt.figure(figsize=(2,2))

# Put a line inside the vector field.
lw = 8 # line-width
pos = 0.5
plt.plot([0,0],[-pos,+pos],color='gray',linewidth=lw,alpha=0.4)

# Make a uniform vector field.
step = 0.25
x,y = np.meshgrid(np.arange(-1,1+step,step),
                  np.arange(-1,1+step,step),
                  indexing='ij')
plt.quiver(x,y,1,0,angles='xy',scale_units='xy',color='k')

plt.axis('square')
plt.axis('off')
plt.xlim(np.array([-1,1])*1.1)
plt.ylim(np.array([-1,1])*1.1)
plt.savefig('fig_ch3_uniform_flux_through_line.pdf')
plt.show()
```

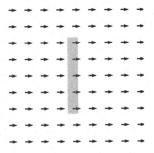

Figure 3.4

Not surprisingly, the amount of vector field's flow is proportional to the size or extent of the boundary, as well as the magnitude of the vectors at the boundary, as illustrated by the following figures.

```
# Code Block 3.5

scale = 3
N = 8

fig = plt.figure(figsize=(6,3))
for i in range(3):
    vec_mag = 1
    lim = 2**(i-2)
    step = 0.1

    plt.subplot(1,4,i+1)
    y = np.linspace(-lim,lim,N)
    y = np.arange(-lim,lim+step,step)
    x = np.zeros(len(y))
    plt.quiver(x,y,vec_mag,0,color='k',
                angles='xy',scale_units='xy',scale=scale)
    plt.plot([0,0],[-lim,+lim],color='gray',linewidth=lw,alpha=0.4)
    plt.axis('square')
    plt.axis('off')
    plt.xlim(np.array([-0.5,1])*0.8)
    plt.ylim(np.array([-1,1])*1.1)
    plt.title('L = %2.1f'%(2*lim))

plt.tight_layout()
plt.savefig('fig_ch3_diff_boundary_extent.pdf',bbox_inches='tight')
plt.show()

fig = plt.figure(figsize=(6,3))
N = 11
for i in range(3):
    vec_mag = 2**(i-1) # (0.5, 1, 2)
    plt.subplot(1,4,i+1)
    plt.quiver(np.zeros(N),np.linspace(-1,1,N)*0.5,vec_mag,0,
                color='k',angles='xy',scale_units='xy',scale=scale)
    plt.plot([0,0],[-0.5,+0.5],color='gray',linewidth=lw,alpha=0.4)
    plt.axis('square')
    plt.axis('off')
    plt.xlim(np.array([-0.5,1])*0.8)
    plt.ylim(np.array([-1,1])*1.1)
    plt.title('|v| = %2.1f'%vec_mag)

plt.tight_layout()
plt.savefig('fig_ch3_diff_v_mag.pdf',bbox_inches='tight')
plt.show()
```

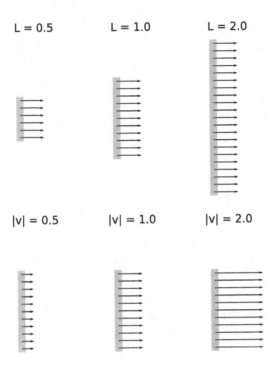

Figure 3.5

Given the above observations, a reasonable initial expression for quantifying the amount of flow would be an integral of vector magnitudes over the range of a boundary, $\int_a^b |\vec{v}|dl$, where $\vec{v}$ is the vector at the boundary, and the total length L of the boundary is broken up into infinitesimal segments dl, such that $\int_a^b dl = L$.

Let's sharpen this idea further. How should we deal with the case when the flow and the boundary are not perpendicular to each other? Consider the following two cases. In the first case, the boundary and the vectors are perpendicular. In the second case, the boundary is slanted (non-orthogonal) relative to the vectors. Even though the slanted boundary is longer, both boundaries span the same vertical range, and the amounts of flow would be equal. Imagine an analogous situation where a water hose is connected to a faucet that delivers a fixed amount of water for a given time. Whether the end of the hose is straight (so that its cross-section is perpendicular to the flow of water) or whether the hose end has a slanted cut, the amount of water outpouring will be the same.

```
# Code Block 3.6

lw = 8
pos = 0.5

# Make a uniform vector field.
step = 0.25
x,y = np.meshgrid(np.arange(-1,1+step,step),
                  np.arange(-1,1+step,step),
                  indexing='ij')

fig = plt.figure(figsize=(6,3))

plt.subplot(1,2,1)
plt.title('(a) Perpendicular')
plt.plot([+0.0,-0.0],[-0.5,+0.5],color='gray',linewidth=lw,alpha=0.4)

plt.subplot(1,2,2)
plt.title('(b) Slanted')
plt.plot([+0.6,-0.6],[-0.5,+0.5],color='gray',linewidth=lw,alpha=0.4)

for i in range(2):
    plt.subplot(1,2,i+1)
    plt.quiver(x,y,1,0,angles='xy',scale_units='xy',color='k')
    plt.axis('square')
    plt.axis('off')
    plt.xlim(np.array([-1,1])*1.1)
    plt.ylim(np.array([-1,1])*1.1)

plt.tight_layout()
plt.savefig('fig_ch3_boundary_slanted.pdf',bbox_inches='tight')
plt.show()
```

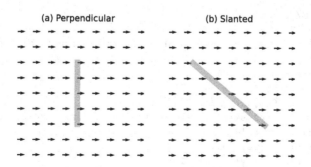

Figure 3.6

This observation motivates us to improve the earlier definition for the amount of flow by introducing a unit normal vector $\hat{n}$ that points perpendicular to the boundary. Note that $\vec{v}$ at the boundary has two components that are orthogonal to each other. $\vec{v}_\perp$ is perpendicular to the boundary and parallel to $\hat{n}$. $\vec{v}_\parallel$ is parallel to the boundary and perpendicular to $\hat{n}$. Therefore, $\vec{v}_\perp \cdot \vec{v}_\parallel = 0$. $\vec{v}_\parallel$ does not contribute to the flow through the boundary. Only $\vec{v}_\perp$ does, and $|\vec{v} \cdot \hat{n}| = |\vec{v}_\perp|$.

Hence, we have an improved definition of the quantity of flow, known as the flux and often denoted with the symbol Φ:

$$\Phi = \int_a^b (\vec{v} \cdot \hat{n})\, dl.$$

3.3 FLUX CALCULATION

In this section, we will build a computational routine for calculating flux through an arbitrary boundary for a given vector field in two dimensions. First, take a look at a few examples of slanted boundary lines and their normal vectors. Once we have a slope of the boundary line, we can calculate the slope of its normal vector as its negative reciprocal.

```
# Code Block 3.7

lw = 8
theta_range = [-90,-75,-60,-45] # in degrees

fig = plt.figure(figsize=(6,3))
for i,theta in enumerate(theta_range):
    plt.subplot(1,4,i+1)
    y0 = -0.5
    y1 = +0.5
    x0 = y0/np.tan(theta*np.pi/180)
    x1 = -x0
    plt.plot([x0,x1],[y0,y1],color='gray',linewidth=lw,alpha=0.4)
    nx = 1
    ny = -(x1-x0)/(y1-y0)
    n_mag = np.sqrt(nx**2+ny**2)
    plt.quiver(0,0,nx/n_mag,ny/n_mag,
               angles='xy',scale_units='xy',color='black',scale=2)
    plt.axis('square')
    plt.xlim(np.array([-1,1])*1.1)
    plt.ylim(np.array([-1,1])*1.1)
    plt.xticks((-1,0,1))
```

```
   plt.yticks((-1,0,1))

plt.tight_layout()
plt.savefig('fig_ch3_normal_vectors.pdf',bbox_inches='tight')
plt.show()
```

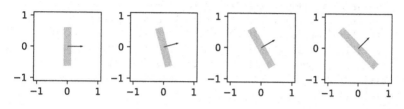

Figure 3.7

If there is a curved boundary, it can be divided into many short, approximately linear segments, and a normal vector at each segment can be found. More specifically, consider a set of coordinates that describe a continuous boundary: $(x_1, y_1), (x_2, y_2), \cdots, (x_N, y_N)$. A normal vector $\hat{n}$ at point (x_i, y_i) can be approximated by finding a line perpendicular to a line between (x_{i-1}, y_{i-1}) and (x_{i+1}, y_{i+1}). $\hat{n}$ has a slope of $-\frac{x_{i+1}-x_{i-1}}{y_{i+1}-y_{i-1}}$ at a point (x_i, y_i). The following code block demonstrates this idea.

```
# Code Block 3.8

# Illustration of how a normal vector is found.

# Define a boundary.
step = 0.05
x = np.arange(0,1,step)
y = np.sqrt(1**2 - x**2)

i = 16 # Point to focus on.

plt.figure(figsize=(4,6))
plt.subplot(3,1,1)

plt.scatter(x,y,color='gray')
plt.plot(x,y,color='gray')
plt.text(x[i+0]+0.05,y[i+0],r"$(x_{i}, y_{i})$")
plt.axis('square')
plt.xlim((0.0,1.2))
plt.ylim((0.0,1.2))
plt.xlabel('x')
plt.ylabel('y')
```

```
plt.xticks((0,0.5,1))
plt.yticks((0,0.5,1))
plt.title('Boundary')

space = 0.012
plt.subplot(3,1,2)
plt.scatter(x[i-1:i+2],y[i-1:i+2],color='gray')
plt.plot([x[i-1],x[i+1]],[y[i-1],y[i+1]],color='gray')
plt.text(x[i+0]+space,y[i+0],r"$(x_{i}, y_{i})$")
plt.text(x[i-1]+space,y[i-1],r"$(x_{i-1}, y_{i-1})$")
plt.text(x[i+1]+space,y[i+1],r"$(x_{i+1}, y_{i+1})$")
plt.axis('square')
plt.xlim((0.7,1.0))
plt.ylim((0.5,0.7))
plt.xlabel('x')
plt.ylabel('y')
plt.xticks((0.7,0.8,0.9))
plt.yticks((0.5,0.6,0.7))
plt.title('Line between neighbors')

plt.subplot(3,1,3)

# Calculate the slope of two immediate neighbors.
slope = (y[i+1]-y[i-1])/(x[i+1]-x[i-1])
# Calcualte the slope of an orthogonal line.
norm_vec_slope = -1/(slope)
# Find the components of the normal vector.
u, v = 1, norm_vec_slope
# Normalize the vector.
mag = np.sqrt(u**2+v**2)
u, v = u/mag, v/mag

plt.scatter(x[i-1:i+2],y[i-1:i+2],color='gray')
plt.quiver(x[i],y[i],u,v,color='black',
           angles='xy',scale_units='xy',scale=10,width=0.01)
plt.text(x[i+0]+space,y[i+0]-0.01,r"$(x_{i}, y_{i})$")
plt.axis('square')
plt.xlabel('x')
plt.ylabel('y')
plt.xlim((0.7,1.0))
plt.ylim((0.5,0.7))
plt.xticks((0.7,0.8,0.9))
plt.yticks((0.5,0.6,0.7))
plt.title('Normal Vector')

plt.tight_layout()
plt.savefig('fig_ch3_normal_illustrate.pdf',bbox_inches='tight')
plt.show()
```

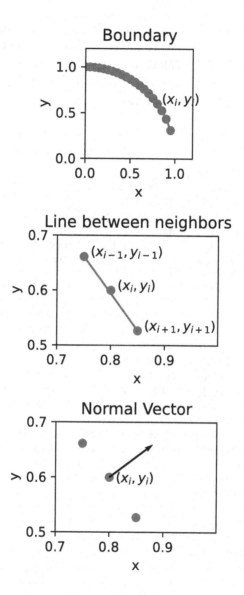

Figure 3.8

In the following code block, we package this idea into a reusable function
`get_normals()`, which accepts as an input argument, a set of adjacent
points that defines a boundary. Note that x and y denote the x and
y coordinates of the boundary. The indexing convention in the code is
such that if x has N elements, x[:-2] refers to the first $N-2$ elements

(without the last two elements), and x[2:] refers to the last $N-2$ elements, starting from the third element (without the first two elements). Therefore, x[2:]-x[:-2] calculates $x_{i+1} - x_{i-1}$ between the two edge points (x[0] and x[-1]) in the boundary. The edge points have only one immediate neighbor, so the normal vectors are not calculated at those positions.

```python
# Code Block 3.9

def get_normals (boundary):
    # The input argument defines a boundary
    # as a set of adjacent points.
    very_small_num = 10**(-10) # avoid divide by zero.
    x, y = boundary[0], boundary[1]
    slope = (y[2:]-y[:-2])/(x[2:]-x[:-2] + very_small_num)
    norm_vec_slope = -1/(slope + very_small_num)
    u, v = 1, norm_vec_slope
    mag = np.sqrt(u**2+v**2)
    u, v = u/mag, v/mag
    n = np.vstack((u,v))
    return n

def plot_normals (boundary,ax,scale=2):
    n = get_normals(boundary)
    x, y = boundary[0], boundary[1]
    ax.scatter(x,y,color='gray')
    ax.plot(x,y,color='gray')
    ax.quiver(x[1:-1],y[1:-1],n[0],n[1],color='gray',
              angles='xy',scale_units='xy',scale=scale)
    ax.axis('equal')
    ax.axis('square')
    #ax.set_xlabel('x')
    #ax.set_ylabel('y')
    ax.set_xlim((-1.0,2.0))
    ax.set_ylim((-1.5,1.5))
    ax.set_xticks((-1,0,1,2))
    ax.set_yticks((-1,0,1))
    return

# Examples of normal vectors for different boundaries.
fig, ax = plt.subplots(1,4,figsize=(6,3),sharey=True)

step = 0.2
y = np.arange(-1,1+step,step)

# Vertical line.
x = np.zeros(len(y))
p0 = np.vstack((x,y))
```

```
plot_normals(p0,ax[0])

# Slanted line
x = -y/3
p1 = np.vstack((x,y))
plot_normals(p1,ax[1])

# Parabola (concave from right)
x = 1-y**2
p2 = np.vstack((x,y))
plot_normals(p2,ax[2])

# Half circle.
theta = np.arange(np.pi/2,-np.pi/2-step,-step)
x = np.cos(theta)
y = np.sin(theta)
p3 = np.vstack((x,y))
plot_normals(p3,ax[3])
plt.tight_layout()
plt.savefig('fig_ch3_normal_diff_boundary.pdf',bbox_inches='tight')
plt.show()
```

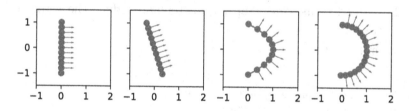

Figure 3.9

Now that we have a function for calculating the normal vectors along a boundary, let's develop a function for calculating the flux for a given boundary and a vector field. As shown in the following code block, the total flux is calculated by adding up the dot products between the normal vector $\hat{n}$ (n[0] and n[1]) and the vector $\vec{v}$ (xv and yv).

The infinitesimal length element dl is calculated by finding the average distance to the two immediate neighbors. In other words, $dl_i = \frac{1}{2}(\sqrt{(x_{i+1} - x_i)^2 + (y_{i+1} - y_i)^2} + \sqrt{(x_i - x_{i-1})^2 + (y_i - y_{i-1})^2})$. x[1:] refers to the x-coordinates of the boundary, starting from the second element to the last element in the array, and x[:-1] refers to the x-coordinates of the boundary, starting from the first element to the second to the last element in the array. Hence, `x[1:]-x[:-1]` produces an array of differences between the adjacent x-coordinate values.

```
# Code Block 3.10

def get_flux(boundary,vfield):
    x, y = boundary[0], boundary[1]
    dl_neighbor = np.sqrt((x[1:]-x[:-1])**2+(y[1:]-y[:-1])**2)
    dl = (0.5)*(dl_neighbor[1:]+dl_neighbor[:-1])
    n = get_normals(boundary)
    xv, yv = vfield[0][1:-1], vfield[1][1:-1]
    dotprod = xv*n[0] + yv*n[1]
    flux = np.sum(dl*dotprod)
    return flux
```

In the following code block, we put `get_flux()` function to the test. Let's calculate the flux of a uniform vector field with four different boundaries, as shown in the above plot. In all cases, the boundaries span the same vertical range of 2 (between -1 and 1), and the uniform field has a magnitude of 1. Therefore, the total flux is expected to be 2, regardless of the shape of the boundary. The calculation yields a value approaching 2, as we decrease the `step` parameter and reduce the errors from discretization.

```
# Code Block 3.11

step = 0.01
def flux_calculation_example (step=0.01):
    y = np.arange(-1,1+step,step)

    x = np.zeros(len(y))
    p0 = np.vstack((x,y))
    vf0 = np.vstack((np.ones(len(x)),np.zeros(len(y))))

    x = -y/3
    p1 = np.vstack((x,y))
    vf1 = np.vstack((np.ones(len(x)),np.zeros(len(y))))

    x = 1-y**2
    p2 = np.vstack((x,y))
    vf2 = np.vstack((np.ones(len(x)),np.zeros(len(y))))

    theta = np.arange(np.pi/2,-np.pi/2-step,-step)
    p3 = np.vstack((np.cos(theta),np.sin(theta)))
    vf3 = np.vstack((np.ones(len(theta)),np.zeros(len(theta))))

    print('Flux with different boundaries (Answer = 2.0)')
```

```
      print('  Vertical Line %8.7f'%get_flux(p0,vf0))
      print('  Slanted Line  %8.7f'%get_flux(p1,vf1))
      print('  Parabola      %8.7f'%get_flux(p2,vf2))
      print('  Half Circle   %8.7f'%get_flux(p3,vf3))

print('\nstep = 0.1')
flux_calculation_example(step=0.1)

print('\nstep = 0.01')
flux_calculation_example(step=0.01)

print('\nstep = 0.001')
flux_calculation_example(step=0.001)
```

```
step = 0.1
Flux with different boundaries (Answer = 2.0)
  Vertical Line 1.9000000
  Slanted Line  1.9000000
  Parabola      1.9037454
  Half Circle   1.9987149

step = 0.01
Flux with different boundaries (Answer = 2.0)
  Vertical Line 1.9900000
  Slanted Line  1.9900000
  Parabola      1.9900377
  Half Circle   1.9999817

step = 0.001
Flux with different boundaries (Answer = 2.0)
  Vertical Line 1.9990000
  Slanted Line  1.9990000
  Parabola      1.9990004
  Half Circle   1.9999999
```

We observed that with a uniform vector field, the shape of a boundary does not change the total flux, as long as the endpoints of a boundary are fixed, but this was such a simplistic scenario. Let's continue our exploration of the flux of a vector field.

3.4   FLUX THROUGH AN ENCLOSURE

Let's consider a boundary whose starting and end points meet. Such a boundary makes an enclosed shape like a square or a circle. The code block below presents three different vector fields with a square boundary.

Continuing with the analogy that a vector field is like water flow, in the first case, more water flows into the left side of the square boundary than the outflow from the right side. In the second case, there is a uniform water flow where the influx and outflux of water are equal. In the third case, more water leaves out of the right side of the square. Because the vectors are parallel to the top and bottom sides of the square boundary, no water is flowing into or out of the square region from those two sides. Inside the square region, we would see water accumulating (if influx > outflux), staying at a constant level (if influx = outflux), or drying out (if influx < outflux).

```
# Code Block 3.12

scale = 8
step = 0.25
x,y = np.meshgrid(np.arange(-1,1+step,step),
                  np.arange(-1,1+step,step),
                  indexing='ij')

pos = 0.5
x0, y0 = 0, pos
x_square = [x0,+pos,+pos,-pos,-pos,x0]
y_square = [y0,+pos,-pos,-pos,+pos,y0]

f, axes = plt.subplots(1,3,figsize=(6,3))

ax = axes[0]
ax.set_title('(a) influx > outflux')
ax.quiver(x,y,1-x,np.zeros(x.shape),
          angles='xy',scale_units='xy',scale=scale,color='k')

ax = axes[1]
ax.set_title('(b) influx = outflux')
ax.quiver(x,y,np.zeros(x.shape)+1,np.zeros(x.shape),
          angles='xy',scale_units='xy',scale=scale,color='k')

ax = axes[2]
ax.set_title('(c) influx < outflux')
ax.quiver(x,y,1+x,np.zeros(x.shape),
          angles='xy',scale_units='xy',scale=scale,color='k')

for i in range(3):
    ax = axes[i]
    ax.plot(x_square,y_square,color='gray',
```

```
            linewidth=lw,alpha=0.4)
    ax.axis('square')
    ax.axis('off')
    ax.set_xlim(np.array([-1,1])*1.2)
    ax.set_ylim(np.array([-1,1])*1.2)

plt.savefig('fig_ch3_flux_square_examples.pdf',bbox_inches='tight')
plt.show()
```

(a) influx > outflux (b) influx = outflux (c) influx < outflux

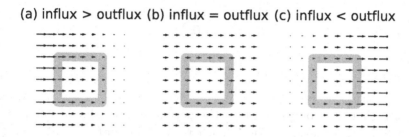

Figure 3.10

Like this example, we often want to analyze the total amount of flux into or out of an enclosed region. In the following several code blocks, we update some of the above codes and build up new ones for calculating flux through an enclosure. In order to demonstrate why some additional codes are necessary, let's build a circular boundary and apply the plot_normals() function we developed previously. We quickly notice that some normal vectors point into the center of the circle and others point outward because, for a given boundary segment, there are two possible perpendicular directions. We need to pick a direction and resolve this inherent ambiguity.

```
# Code Block 3.13

step = np.pi/10
r = 0.75
phi = np.arange(0,2*np.pi,step)
p = np.vstack((r*np.cos(phi),r*np.sin(phi)))
fig, ax = plt.subplots(1,1,figsize=(2,2))
plot_normals(p,ax)
plt.xlim((-1.5,1.5))
plt.ylim((-1.5,1.5))
plt.savefig('fig_ch3_normal_in_out.pdf',bbox_inches='tight')
```

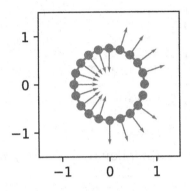

Figure 3.11

We will choose a convention that the normal vectors should point out-ward from an enclosure. The following functions, `get_normals_enc()` and `plot_normals_enc()`, accept an additional input argument `inside` that specifies a point that disambiguates the inside and outside of a boundary. The default `inside` point is the origin $(0,0)$ of the coordinate system. We also make a simplifying assumption that the boundary surface is not too complex (i.e., no dips or concavity).

Then, consider a vector between a point on the boundary $\vec{p}_b$ and the inside point $\vec{p}_i$. We can define a vector that points outward by $\vec{o} = \vec{p}_b - \vec{p}_i = (x_b - x_i, y_b - y_i)$. Given a normal vector $\hat{n}$ on the boundary, we can orient it outward by multiplying it by the sign of a dot product between $\vec{o}$ and $\hat{n}$. The sign of the dot product will be positive if $\hat{n}$ is indeed pointing outward. If not, the direction of $\hat{n}$ can be reversed. This strategy is implemented in `get_normals_enc()`.

There is another subtle difference between `get_normals()` and `get_normals_enc()`. Because the normal vector at a particular point is calculated based on two neighboring points, the normal vector can not be calculated at the first or the last points in the boundary. This omission is apparent in the above figure. When we are working with an enclosed boundary, the first and the last points are adjacent to each other, and the normal vectors can be calculated for all points on the boundary by concatenating the first two points at the end of the boundary array (`boundary_ext = np.hstack((boundary,boundary[:,:2]))` in the code). The following plot shows that the normal vectors are displayed for all points along the boundary.

```
# Code Block 3.14

def get_normals_enc (boundary,inside=(0,0)):
    boundary_ext = np.hstack((boundary,boundary[:,:2]))
    x, y = boundary_ext[0], boundary_ext[1]

    very_small_num = 10**(-10) # avoid divide by zero.
    slope = (y[2:]-y[:-2])/(x[2:]-x[:-2] + very_small_num)
    norm_vec_slope = -1/(slope + very_small_num)
    u, v = 1, norm_vec_slope
    mag = np.sqrt(u**2+v**2)
    u, v = u/mag, v/mag
    n = np.vstack((u,v))

    # Calculate the sign of a dot product between n and
    # vector between (x,y)-inside, which determines
    # which way n should point.
    x, y = x[1:-1]-inside[0], y[1:-1]-inside[1]
    dot_prod_sign = np.sign(x*n[0] + y*n[1])
    n = n*dot_prod_sign
    return n

def plot_normals_enc (boundary,ax,inside=(0,0),scale=3):
    n = get_normals_enc(boundary,inside=inside)

    boundary_ext = np.hstack((boundary,boundary[:,:2]))
    x, y = boundary_ext[0], boundary_ext[1]

    color = '#CCCCCC'
    ax.scatter(x,y,color='gray')
    ax.plot(x,y,color='gray')
    # Mark the inside point with a diamond marker.
    #ax.scatter(inside[0],inside[1],marker='d',color='black')
    ax.quiver(x[1:-1],y[1:-1],n[0],n[1],color=color,
              angles='xy',scale_units='xy',scale=scale)
    ax.set_xlabel('x')
    ax.set_ylabel('y')
    ax.axis('equal')
    ax.axis('square')
    return

fig, ax = plt.subplots(1,1,figsize=(2,2))
plot_normals_enc(p,ax,inside=(0,0))
lim = 1.5
ax.set_xlim((-lim,lim))
ax.set_ylim((-lim,lim))
ax.set_xticks((-1,0,1))
ax.set_yticks((-1,0,1))
```

```
plt.savefig('fig_ch3_normal_outward.pdf',bbox_inches='tight')
plt.show()
```

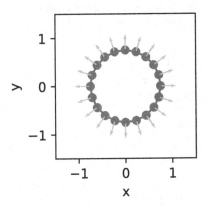

Figure 3.12

Throughout our study of electrodynamics, we will be considering differently shaped enclosures. Sometimes, we will show that certain mathematical quantities about a vector field are independent of the shape of the enclosing boundary. For example, the flux of a special vector field, whose magnitude falls off as $\frac{1}{r}$ in two dimensions, is the same whether the enclosing boundary is a square or a circle, as long as certain conditions hold (we will discuss what these conditions are in a later chapter). In some other examples, we will arrange the sources of a vector field (e.g., electric charges or currents) in different ways. Therefore, in the following code block, we define two functions we will use multiple times throughout the book. The two helper functions, `points_along_circle()` and `points_along_square()`, return a set of points that trace circular and square enclosures. Both functions have a parameter `step` that controls how closely the adjacent points are spaced.

```
# Code Block 3.15

def points_along_circle (r=1, step=np.pi/10):
    phi = np.arange(-np.pi,np.pi,step)
    p = np.vstack((r*np.cos(phi),r*np.sin(phi)))
    return p

def points_along_square (s=1, step=0.1):
    # Get points along a square, whose sides are of length s.
    d = s/2
```

```
    one_side = np.arange(-d,d,step)
    N = len(one_side)
    p_top = np.vstack((+one_side,np.zeros(N)+d)) # top side
    p_rgt = np.vstack((np.zeros(N)+d,-one_side)) # right side
    p_bot = np.vstack((-one_side,np.zeros(N)-d)) # bottom side
    p_lft = np.vstack((np.zeros(N)-d,+one_side)) # left side
    p = np.hstack((p_top,p_rgt,p_bot,p_lft))
    return p

fig = plt.figure(figsize=(2,2))
p = points_along_square(s=4)
plt.plot(p[0],p[1],color='gray')
p = points_along_circle(r=2,step=np.pi/100)
plt.plot(p[0],p[1],color='gray')
p = points_along_circle(r=np.sqrt(8),step=np.pi/100)
plt.plot(p[0],p[1],color='gray')

plt.axis('equal')
plt.axis('square')
plt.xlabel('x')
plt.ylabel('y')
plt.xlim((-4,4))
plt.ylim((-4,4))
plt.xticks((-2,0,2))
plt.yticks((-2,0,2))
plt.savefig('fig_ch3_diff_enclosures.pdf',bbox_inches='tight')
plt.show()
```

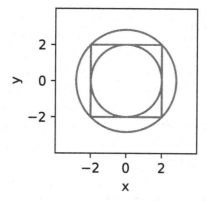

Figure 3.13

In addition to the above functions, let's create two more helper functions, get_vfield_uniform() and get_vfield_radial(), which return uniform and radial vector fields, respectively. Both functions accept an input argument p, a set of points at which the vector fields are to be specified.

The get_vfield_radial() function calculates the distance from the origin as r = np.sqrt(np.sum(p**2,axis=0)), which computes $r = \sqrt{x^2 + y^2}$. In order to generate a vector field whose magnitude decreases as $\frac{1}{r}$, we must divide x and y coordinates by r^2. To generate a vector field whose magnitude is independent of r, we divide x and y by r. Whenever there is a division, we could run into an unfortunate situation where the divisor (in this case r) is too small and causes serious numerical errors. There are many different ways of avoiding division by zero or small numbers. Here, we drop points that are too close to the origin. What is "too close"? In our case, we drop the points whose distance from the origin is less than 1 percent of the maximum distance. We find indices that satisfy this condition (valid_idx = np.where(r > np.max(r)*0.01) in the code) and squeeze out the unwanted points with the squeeze() method. As a result, the get_vfield_radial() function may return p that contains the subset of the original input argument p.

```
# Code Block 3.16

def get_vfield_uniform (p,type='right'):
    if type=='right':
        dx, dy = 1, 0
    if type=='left':
        dx, dy = -1, 0
    if type=='up':
        dx, dy = 0, 1
    if type=='down':
        dx, dy = 0, -1
    vf = np.zeros(p.shape)
    vf[0] = dx
    vf[1] = dy
    return vf, p

def get_vfield_radial (p,type='1/r'):

    r = np.sqrt(np.sum(p**2,axis=0))
    # The above line is the same as:
    #r = np.sqrt(p[0]**2+p[1]**2)

    # When we are dividing by r, a small value of r
```

```
    # can blow up the magnitude of a vector.
    # There are different ways of avoiding this situation.
    # Here, we will drop points with relatively small r
    # (small compared to maximum r).
    # Therefore, this function may return p that is
    # different from the original input argument p.
    valid_idx = np.where(r > np.max(r)*0.01)
    q, s = p[:,valid_idx].squeeze(), r[valid_idx].squeeze()
    p, r = q, s

    if type=='r':
        dx, dy = p[0], p[1]
    if type=='1':
        dx, dy = p[0]/r, p[1]/r
    if type=='1/r':
        dx, dy = p[0]/r**2, p[1]/r**2
    if type=='1/r**2':
        dx, dy = p[0]/r**3, p[1]/r**3

    vf = np.zeros(p.shape)
    vf[0], vf[1] = dx, dy
    return vf, p

def plot_vfield (p,vf):
    fig = plt.figure(figsize=(2.5,2.5))
    plt.quiver(p[0],p[1],vf[0],vf[1],
               angles='xy',scale_units='xy')
    plt.axis('equal')
    plt.axis('square')
    plt.xlabel('x')
    plt.ylabel('y')
    lim = 1.25
    plt.xlim((-lim,lim))
    plt.ylim((-lim,lim))
    plt.xticks((-1,0,1))
    plt.yticks((-1,0,1))

step = 0.2
x,y = np.meshgrid(np.arange(-1,1+step,step),
                  np.arange(-1,1+step,step))
p = np.vstack((x.flatten(),y.flatten()))

vf, p = get_vfield_uniform(p,type='down')
plot_vfield(p,vf)
plt.title('Uniform Vector Field: Down')
plt.savefig('fig_ch3_diff_vfield_uniform_down.pdf')
plt.show()

vf, p = get_vfield_uniform(p,type='left')
```

```
plot_vfield(p,vf)
plt.title('Uniform Vector Field: Left')
plt.savefig('fig_ch3_diff_vfield_uniform_left.pdf')
plt.show()

vf, p = get_vfield_radial(p,type='r')
plot_vfield(p,vf)
plt.title('Radial Vector Field: r')
plt.savefig('fig_ch3_diff_vfield_linear_r.pdf')
plt.show()

vf, p = get_vfield_radial(p,type='1')
plot_vfield(p,vf)
plt.title('Radial Vector Field: Constant')
plt.savefig('fig_ch3_diff_vfield_constant_r.pdf')
plt.show()

vf, p = get_vfield_radial(p,type='1/r')
plot_vfield(p,vf)
plt.title('Radial Vector Field: 1/r')
plt.savefig('fig_ch3_diff_vfield_inverse_r.pdf')
plt.show()

vf, p = get_vfield_radial(p,type='1/r**2')
plot_vfield(p,vf)
plt.title('Radial Vector Field: 1/r^2')
plt.savefig('fig_ch3_diff_vfield_inverse_r2.pdf')
plt.show()
```

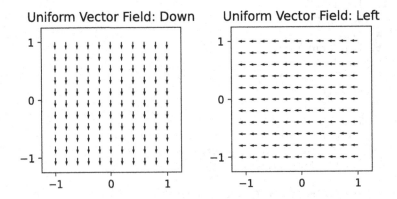

Figure 3.14

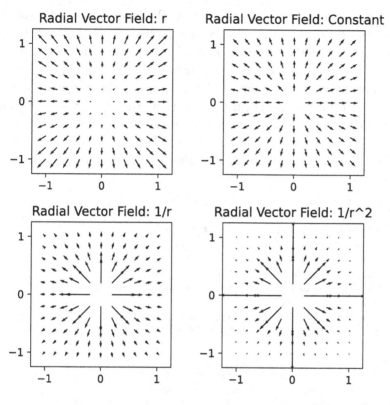

Figure 3.15

Let's also upgrade the get_flux() function to work with an enclosure. We will call it get_flux_enc().

```
# Code Block 3.17

def get_flux_enc (boundary,vfield,inside=(0,0)):
    boundary_ext = np.hstack((boundary,boundary[:,:2]))
    x, y = boundary_ext[0], boundary_ext[1]
    dl_neighbor = np.sqrt((x[1:]-x[:-1])**2+(y[1:]-y[:-1])**2)
    dl = (0.5)*(dl_neighbor[1:]+dl_neighbor[:-1])
    n = get_normals_enc(boundary,inside=inside)
    xv, yv = vfield[0], vfield[1]
    dotprod = xv*n[0] + yv*n[1]
    flux = np.sum(dl*dotprod)
    return flux
```

Let's combine these helper functions to calculate the total flux through different boundary shapes: circles and squares. For a uniform vector field, we have already demonstrated that the flux through different open boundaries (such as a vertical line, slanted line, parabola, or half circle) is equal as long as their endpoints are identical. Thus, it is not surprising that the total flux of a uniform vector field through an enclosed boundary is always zero. The following figures display the normal vectors (in gray) and the vector fields (in black). Along the left side of the boundary, the dot products between the normal vectors and the vector field produce negative values, while the dot products are positive on the right side of the boundary. When these values are added together, the total flux is zero.

```python
# Code Block 3.18

fig, axs = plt.subplots(1,2,figsize=(6,3))

ax = axs[0]
p = points_along_square(s=2,step=0.25)
vf, p = get_vfield_uniform (p,type='right')
plot_normals_enc(p,ax)
ax.quiver(p[0],p[1],vf[0],vf[1],color='black')
ax.set_title('Flux = %3.2f'%get_flux_enc(p,vf))

ax = axs[1]
p = points_along_circle(r=1,step=np.pi/10)
vf, p = get_vfield_uniform (p,type='right')
plot_normals_enc(p,ax)
ax.quiver(p[0],p[1],vf[0],vf[1],color='black')
ax.set_title('Flux = %3.2f'%get_flux_enc(p,vf))

for i in range(2):
    ax = axs[i]
    lim = 2
    ax.set_xlim((-lim,lim))
    ax.set_ylim((-lim,lim))
    ax.set_xticks((-1,0,1))
    ax.set_yticks((-1,0,1))

plt.tight_layout()
plt.savefig('fig_ch3_uniform_v_flux.pdf')
plt.show()
```

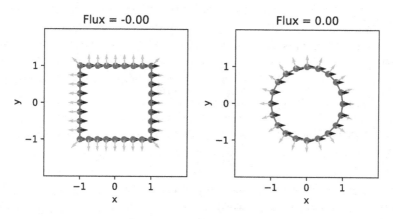

Figure 3.16

The accuracy of the numerical calculation for flux improves as more points on the boundary are considered, as shown below.

```
# Code Block 3.19

print('Flux should approach zero with smaller step size.')
for step in (0.5,0.1,0.05,0.01):
    p = points_along_circle(r=1,step=step)
    vf, p = get_vfield_uniform(p,type='right')
    print('With step = %6.5f, flux = %9.8f'%(step,get_flux_enc(p,vf)))
```

```
Flux should approach zero with smaller step size.
With step = 0.50000, flux = 0.01032620
With step = 0.10000, flux = 0.00003844
With step = 0.05000, flux = 0.00000874
With step = 0.01000, flux = 0.00000011
```

3.5   A SPECIAL VECTOR FIELD

We conclude this chapter by considering a special vector field that points radially with its magnitude inversely proportional to r in two dimensions. As shown by the following calculations, the flux of this radial vector field does not depend on the shape or size of the boundary. The flux is close to **6.29** or 2π.

```
# Code Block 3.20

# The flux of the 1/r vector field is constant,
# no matter what.
```

```
step = 0.01

fig = plt.figure(figsize=(3,3))
# Circular boundary
p_cl = points_along_circle(r=3,step=step)
vf_cl, p_cl = get_vfield_radial (p_cl,type='1/r')
plt.plot(p_cl[0],p_cl[1],color='gray')

p_cm = points_along_circle(r=2,step=step)
vf_cm, p_cm = get_vfield_radial (p_cm,type='1/r')
plt.plot(p_cm[0],p_cm[1],color='gray')

p_cs = points_along_circle(r=1,step=step)
vf_cs, p_cs = get_vfield_radial (p_cs,type='1/r')
plt.plot(p_cs[0],p_cs[1],color='gray')

# Square boundary
p_sl = points_along_square(s=6,step=step)
vf_sl, p_sl = get_vfield_radial (p_sl,type='1/r')
plt.plot(p_sl[0],p_sl[1],color='gray')

p_sm = points_along_square(s=4,step=step)
vf_sm, p_sm = get_vfield_radial (p_sm,type='1/r')
plt.plot(p_sm[0],p_sm[1],color='gray')

p_ss = points_along_square(s=2,step=step)
vf_ss, p_ss = get_vfield_radial (p_ss,type='1/r')
plt.plot(p_ss[0],p_ss[1],color='gray')

step = 1
x,y = np.meshgrid(np.arange(-5,5+step,step),
                  np.arange(-5,5+step,step))
p = np.vstack((x.flatten(),y.flatten()))
vf, p = get_vfield_radial(p,type='1/r')
plt.quiver(p[0],p[1],vf[0],vf[1],angles='xy',scale_units='xy')

plt.axis('equal')
plt.axis('square')
plt.xlabel('x')
plt.ylabel('y')
plt.xlim((-5,5))
plt.ylim((-5,5))
plt.xticks((-4,-2,0,2,4))
plt.yticks((-4,-2,0,2,4))
plt.title('Different Boundaries under radial vector field')
plt.savefig('fig_ch3_radial_flux.pdf',bbox_inches='tight')
plt.show()

print('Flux with a large  circle = %8.7f'%get_flux_enc(p_cl,vf_cl))
```

```
print('Flux with a medium circle = %8.7f'%get_flux_enc(p_cm,vf_cm))
print('Flux with a small  circle = %8.7f'%get_flux_enc(p_cs,vf_cs))
print('Flux with a large  square = %8.7f'%get_flux_enc(p_sl,vf_sl))
print('Flux with a medium square = %8.7f'%get_flux_enc(p_sm,vf_sm))
print('Flux with a small  square = %8.7f'%get_flux_enc(p_ss,vf_ss))
```

```
Flux with a large  circle = 6.2828454
Flux with a medium circle = 6.2828454
Flux with a small  circle = 6.2828454
Flux with a large  square = 6.2859365
Flux with a medium square = 6.2873043
Flux with a small  square = 6.2913767
```

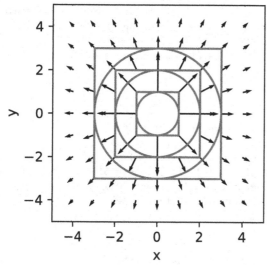

Different Boundaries under radial vector field

Figure 3.17

This constancy is remarkable since, in all cases, the magnitudes and the relative angles between the normals and the vector field are quite different. Four representative cases are shown in the following plot. For drawing the figures, we use relatively large spacings between the points along the boundary, in order to avoid making a messy, cluttered plot. The flux values are calculated with a smaller value of step.

```
# Code Block 3.21

def make_flux_plots(type='1/r', scale=1):
    inside = (0,0)
```

```
fig, axs = plt.subplots(2,2,figsize=(6,6))

ax = axs[0,0]
p = points_along_circle(r=0.6,step=np.pi/12)
vf, p = get_vfield_radial (p,type=type)
plot_normals_enc(p,ax,inside=inside)
ax.quiver(p[0],p[1],vf[0],vf[1],color='black',scale=scale)
p = points_along_circle(r=0.6,step=np.pi/120)
vf, p = get_vfield_radial (p,type=type)
ax.set_title('Flux = %4.3f'%get_flux_enc(p,vf))

ax = axs[0,1]
p = points_along_circle(r=0.3,step=np.pi/12)
vf, p = get_vfield_radial (p,type=type)
plot_normals_enc(p,ax,inside=inside)
ax.quiver(p[0],p[1],vf[0],vf[1],color='black',scale=scale)
p = points_along_circle(r=0.3,step=np.pi/120)
vf, p = get_vfield_radial (p,type=type)
ax.set_title('Flux = %4.3f'%get_flux_enc(p,vf))

step = 0.2

ax = axs[1,0]
p = points_along_square(s=1.6,step=step)
vf, p = get_vfield_radial (p,type=type)
plot_normals_enc(p,ax,inside=inside)
ax.quiver(p[0],p[1],vf[0],vf[1],color='black',scale=scale)
p = points_along_square(s=1.6,step=0.01)
vf, p = get_vfield_radial (p,type=type)
ax.set_title('Flux = %4.3f'%get_flux_enc(p,vf))

ax = axs[1,1]
p = points_along_square(s=1,step=step)
vf, p = get_vfield_radial (p,type=type)
plot_normals_enc(p,ax,inside=inside)
ax.quiver(p[0],p[1],vf[0],vf[1],color='black',scale=scale)
p = points_along_square(s=1,step=0.01)
vf, p = get_vfield_radial (p,type=type)
ax.set_title('Flux = %4.3f'%get_flux_enc(p,vf))

for i in range(2):
    for j in range(2):
        ax = axs[i,j]
        lim = 1.5
        ax.set_xlim((-lim,lim))
        ax.set_ylim((-lim,lim))
        ax.set_xticks((-1,0,1))
        ax.set_yticks((-1,0,1))
```

```
    plt.tight_layout()

    return

print('Vector field: radial with 1/r')
make_flux_plots(type='1/r',scale=16)
plt.savefig('fig_ch3_radial_diff_boundary.pdf')
plt.show()
```

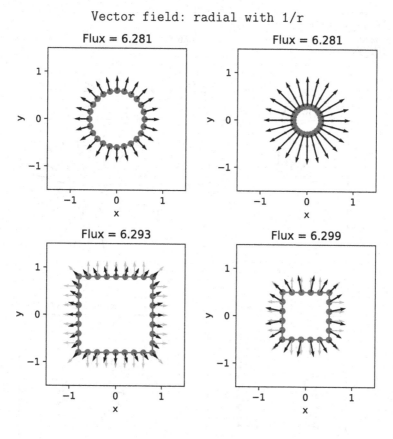

Figure 3.18

In other situations (with vector fields whose magnitudes vary as different powers of r), the flux depends on the shape of the boundary, as shown below.

```
# Code Block 3.22

print('Vector field: radial with 1/r**2')
```

```
make_flux_plots(type='1/r**2',scale=32)
plt.savefig('fig_ch3_other_vf_diff_boundary1.pdf')
plt.show()

print('Vector field: radial with 1')
make_flux_plots(type='1',scale=16)
plt.savefig('fig_ch3_other_vf_diff_boundary2.pdf')
plt.show()

print('Vector field: radial with r')
make_flux_plots(type='r',scale=8)
plt.savefig('fig_ch3_other_vf_diff_boundary3.pdf')
plt.show()
```

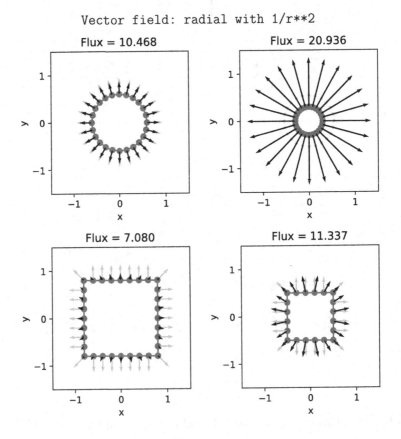

Figure 3.19

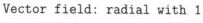

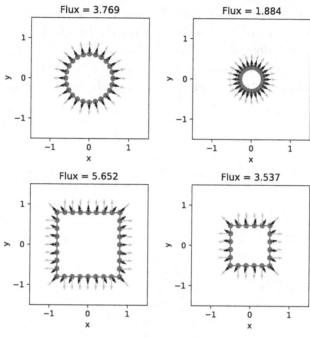

Figure 3.20

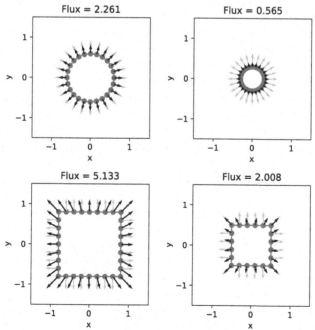

Figure 3.21

In the following chapter, we will discuss why $\frac{1}{r}$ vector field is particularly special in a two-dimensional space. For now, the main takeaway lesson of this chapter is that we can perform interesting mathematical operations on a vector field along a boundary. In the case of a flux calculation, we compute a dot product between the vector field and the normal vectors at each point along the boundary and integrate the dot product values.

## CHAPTER 4

Electric Field

4.1 ELECTRIC CHARGES AND COULOMB'S LAW

At the beginning of the book, we introduced the analogy of water flowing on a landscape. We might say that this vector field that guides the water flow has been created by the abundance or the lack of land mass in the cases of hills and valleys, respectively. In the context of electricity, the vector field of electricity, or electric field, is created by the electric charges. Unlike mass, which is always a non-negative quantity, an electric charge may be positive, neutral, or negative. Canonical examples of electric charges are protons, neutrons, and electrons. In many of our analyses, we will often consider a point charge, which occupies a very small region of space, relative to the length scale of the analysis, so for all practical purposes, it is a point in space. Electric charge is measured in the unit of coulomb (C). The opposite charges (positive and negative) attract each other, and the same charges repel each other. Such interactions are mediated through the electric field.

An electric point charge produces an electric field whose strength decreases as an inverse function of distance away from the charge. More specifically, in two dimensions, if a charge q is placed at position (x_s, y_s), the electric field strength at a point (x, y) is given by:

$$\vec{\mathbf{E}}_{2D}(x, y) = \frac{q}{2\pi\epsilon_0} \left(\frac{x - x_s}{r^2}\hat{\mathbf{x}} + \frac{y - y_s}{r^2}\hat{\mathbf{y}} \right),$$

where $r = \sqrt{(x - x_s)^2 + (y - y_s)^2}$.

DOI: 10.1201/9781003397496-4

Then,

$$|\vec{\mathbf{E}}_{2D}| = \frac{q}{2\pi\epsilon_0}\frac{1}{r},$$

which can be easily verified with $\sqrt{(\frac{x-x_s}{r^2})^2 + (\frac{y-y_s}{r^2})^2} = \frac{1}{r}$. The constant ϵ_0 is known as the permittivity of free space with a value of 8.85×10^{-12} C^2 N m^{-2}. The above formulation is true only in a two-dimensional universe, where the magnitude of the electric field diminishes as $\frac{1}{r}$. We will develop intuitions and formulations about the electric field in two dimensions first because it is easier to visualize. In a later section, we will present a three-dimensional formulation appropriate for our own universe, where the field strength diminishes as $\frac{1}{r^2}$ instead.

The function get_vfield_radial_2d() in the following code block implements the above expression. Given the position of a point charge p_charge, it returns the electric field at positions p. In the code, the constant ϵ_0 is assumed to be one, and a single point charge q is also assumed to have a unitless charge of either $+1$ or -1.

```python
# Code Block 4.1

import numpy as np
import matplotlib.pyplot as plt

def get_vfield_radial_2d (p, p_charge):
    # p: points at which the electric field is calculated.
    # p_charge: location of a point charge.
    x, y = p[0]-p_charge[0], p[1]-p_charge[1]
    r = np.sqrt(x**2 + y**2)

    # Avoid dividing by a very small number.
    valid_idx = np.where(r > np.max(r)*0.01)
    p, r = p[:,valid_idx].squeeze(), r[valid_idx].squeeze()

    vf = np.zeros(p.shape)
    vf[0] = (p[0]-p_charge[0])/r**2
    vf[1] = (p[1]-p_charge[1])/r**2
    vf = vf/(2*np.pi)
    return vf, p
```

Let's visualize the electric field of a point charge placed at the origin. When the sign of the electric charge is negative, we can simply flip the direction of the electric field, as indicated by vf_neg = -vf_pos.

```
# Code Block 4.2

# Set up a grid to plot the vector field.
step = 0.1
x,y = np.meshgrid(np.arange(-1,1+step,step),
                  np.arange(-1,1+step,step))
xs, ys = 0.0, 0.0
p = np.vstack((x.flatten(),y.flatten()))

def tidy_up_ax (ax):
    ax.axis('equal')
    ax.axis('square')
    lim = 1.25
    ax.set_xlim((-lim,lim))
    ax.set_ylim((-lim,lim))
    ax.set_xlabel('x')
    ax.set_ylabel('y')
    ax.set_xticks((-1,0,1))
    ax.set_yticks((-1,0,1))
    return

scale = 6

# Positive point charge
vf_pos, p = get_vfield_radial_2d (p,(xs,ys))
fig = plt.figure(figsize=(2,2))
plt.scatter(xs,ys,marker="+",color='black')
plt.quiver(p[0],p[1],vf_pos[0],vf_pos[1],
           angles='xy',scale_units='xy',scale=scale)
tidy_up_ax(plt.gca())
plt.title('Positive Point Charge')
plt.savefig('fig_ch4_single_charge_pos.pdf',bbox_inches='tight')
plt.show()

# Negative point charge
vf_neg = -vf_pos
fig = plt.figure(figsize=(2,2))
plt.scatter(xs,ys,marker="_",color='black')
plt.quiver(p[0],p[1],vf_neg[0],vf_neg[1],
           angles='xy',scale_units='xy',scale=scale)
tidy_up_ax(plt.gca())
plt.title('Negative Point Charge')
plt.savefig('fig_ch4_single_charge_neg.pdf',bbox_inches='tight')
plt.show()
```

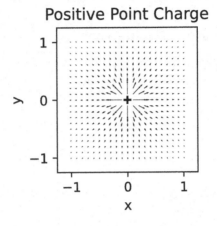

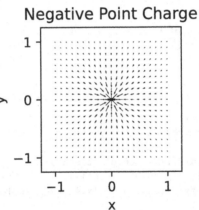

Figure 4.1

4.2 SUPERPOSITION PRINCIPLE

What happens when there is more than one electric charge? The answer is that the electric field simply adds linearly, which is known as the superposition principle. Such an additive property may seem most reasonable and it certainly makes our calculations easier, but it is not a logical necessity. In other contexts and phenomena, the collective "total" may be given by the minimum (e.g., a chain is as strong as its weakest link), average, maximum (e.g., the performance of a group may be determined by the most dominant member), or a nonlinear boost (e.g., the whole may be greater than the sum of its parts). In the case of quantum mechanics, probability amplitudes add linearly ($\psi_1 + \psi_2 + \cdots$), but the observed probability is given by the amplitude squared ($|\psi_1 + \psi_2 + \cdots|^2$).

The laws of electromagnetism seem to operate under the superposition principle, which is an experimental fact. The total electric field can be calculated by adding up individual contributions of electric charges. As shown in the following code block, when there are two point charges, the electric field due to each one can be calculated separately and then added together later (`vf = vf0 + vf1`). The last example is known as a dipole, where a pair of equal and opposite charges are arranged a small distance apart.

```python
# Code Block 4.3

# Two charges.
xs0, ys0 = 0.55, 0.00
xs1, ys1 = -0.55, 0.00
p = np.vstack((x.flatten(),y.flatten()))
vf0, p = get_vfield_radial_2d (p,(xs0,ys0))
vf1, p = get_vfield_radial_2d (p,(xs1,ys1))

# Two positive charges.
vf = vf0 + vf1
fig = plt.figure(figsize=(2,2))
plt.scatter(xs0,ys0,marker="+",color='black')
plt.scatter(xs1,ys1,marker="+",color='black')
plt.quiver(p[0],p[1],vf[0],vf[1],
           angles='xy',scale_units='xy',scale=scale)
tidy_up_ax(plt.gca())
plt.title('Two Positive Charges')
plt.savefig('fig_ch4_double_charge_pos.pdf',bbox_inches='tight')
plt.show()

# Two negative charges.
vf = (-vf0) + (-vf1)
fig = plt.figure(figsize=(2,2))
plt.scatter(xs0,ys0,marker="_",color='black')
plt.scatter(xs1,ys1,marker="_",color='black')
plt.quiver(p[0],p[1],vf[0],vf[1],
           angles='xy',scale_units='xy',scale=scale)
tidy_up_ax(plt.gca())
plt.title('Two Negative Charges')
plt.savefig('fig_ch4_double_charge_neg.pdf',bbox_inches='tight')
plt.show()

vf = vf0 + (-vf1) # Dipole.
fig = plt.figure(figsize=(2,2))
```

```
plt.scatter(xs0,ys0,marker="+",color='black')
plt.scatter(xs1,ys1,marker="_",color='black')
plt.quiver(p[0],p[1],vf[0],vf[1],
          angles='xy',scale_units='xy',scale=scale)
tidy_up_ax(plt.gca())
plt.title('Dipole')
plt.savefig('fig_ch4_double_charge_dipole.pdf',bbox_inches='tight')
plt.show()
```

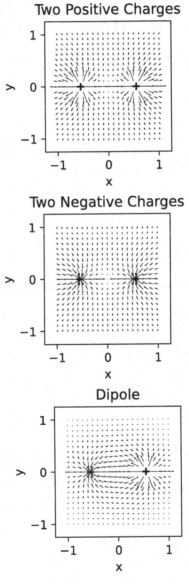

Figure 4.2

4.3   EXAMPLE: INFINITE LINE

Let's apply the superposition principle to calculate the electric field when the charges are distributed along a long line. As a theoretical exercise, consider a line with linear charge density ρ, so that the amount of electric charge along an infinitesimal length dy is given by ρdy. Then, the electric field at a distance D away from this line will be given by:

$$|\vec{E}(x = D)| = \int_{-\infty}^{\infty} dE_x(y),$$

where $dE_x(y)$ is the $\hat{x}$-component of the electric field due to the infinitesimal charge ρdy at position y. We may ignore the $\hat{y}$-component for the following reason. Because we are considering an infinite line, we can choose the point of interest to be at $(D, 0)$ without loss of generality and always find a pair of charges (q_i and q_j in the following figure) that are symmetrically located with respect to the x-axis. Therefore, the $\hat{y}$-components of their electric fields are equal and opposite, while the

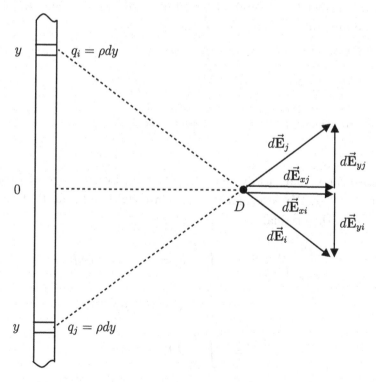

Figure 4.3

$\hat{x}$-components point in the same direction. Thanks to this symmetry, the $\hat{y}$-component of the total field is zero.

The magnitude of the electric field due to an infinitesimal charge at y is:

$$|d\vec{E}(x = D, y)| = \frac{\rho dy}{2\pi\epsilon_0} \frac{1}{\sqrt{D^2 + y^2}}.$$

The x-component of the above expression is

$$|d\vec{E}_x(x = D, y)| = \frac{\rho dy}{2\pi\epsilon_0} \frac{1}{\sqrt{D^2 + y^2}} \frac{D}{\sqrt{D^2 + y^2}} = \frac{\rho dy}{2\pi\epsilon_0} \frac{D}{D^2 + y^2}.$$

Then, the total electric field is

$$|\vec{E}(x = D)| = \frac{\rho}{2\pi\epsilon_0} \int_{-\infty}^{\infty} \frac{D dy}{D^2 + y^2} = \frac{\rho}{2\pi\epsilon_0} \int_{-\infty}^{\infty} \frac{dt}{1 + t^2},$$

where we have defined a new variable $t = y/D$.

The definite integral converges and turns out to be π, as shown by the following code block with the sympy module. The line `sym.Symbol('t')` allows us to treat t as a symbolic constant, and `sym.integrate()` evaluates the integral of a function for the specified range.

```
# Code Block 4.4

import sympy as sym
t = sym.Symbol('t')
# sym.oo is symbolic constant for infinity.
sym.integrate(1/(1+t**2), (t,-sym.oo,sym.oo))
```

π

Let's also show analytically that $\int_{-\infty}^{\infty} \frac{dt}{1+t^2} = \pi$, by a change of variable $t = \tan\theta = \frac{\sin\theta}{\cos\theta}$. We note that $dt = (1 + \tan^2\theta)d\theta$ and that $1 + \tan^2\theta = \frac{1}{\cos^2\theta}$. Then, the original integral becomes

$$\begin{aligned}
\int_{-\infty}^{\infty} \frac{dt}{1 + t^2} &= \int_{-\pi/2}^{\pi/2} \cos^2\theta(1 + \tan^2\theta)d\theta \\
&= \int_{-\pi/2}^{\pi/2} (\cos^2\theta + \sin^2\theta)d\theta \\
&= \int_{-\pi/2}^{\pi/2} d\theta \\
&= \pi.
\end{aligned}$$

We conclude that

$$|\vec{\mathbf{E}}(x = D)| = \frac{\rho}{2\epsilon_0},$$

which is an interesting result, since it is a constant that does not depend on the distance D.

Let's simulate this situation numerically. In the following code, y_max denotes the length of the line. As we increase this length, the simulation result approaches the theoretical limit. Try y_max = 0.5 and y_max = 10. In the latter case, the vectors have almost identical magnitudes, and their directions are horizontal, as expected from the above analysis.

```
# Code Block 4.5

# Infinite line
step = 0.1
x,y = np.meshgrid(np.arange(-1,1+step,step),
                  np.arange(-1,1+step,step))
p = np.vstack((x.flatten(),y.flatten()))
idx = np.where(np.abs(p[0])>0.001)
p = np.vstack((p[0,idx],p[1,idx]))

dy = 0.1
y_max = 10
y_range = np.arange(-y_max,y_max,dy) # Infinite line.

# Test run to see the size of returned p.
vf0, p = get_vfield_radial_2d (p,(0,0))
vf_single = np.zeros(p.shape)

for ys in y_range:
    vf0, p = get_vfield_radial_2d (p,(0,ys))
    vf_single = vf_single + (vf0)*dy

fig = plt.figure(figsize=(2,2))
plt.scatter(np.zeros(y_range.shape),y_range,marker='+',color='black')
plt.quiver(p[0],p[1],vf_single[0],vf_single[1],
           angles='xy',scale_units='xy')
plt.axis('equal')
plt.axis('square')
plt.xlabel('x')
plt.ylabel('y')
lim = 1.0
plt.xlim((-lim,lim))
plt.ylim((-lim,lim))
plt.xticks((-1,0,1))
plt.yticks((-1,0,1))
```

```
plt.savefig('fig_ch4_infinite_line.pdf',bbox_inches='tight')
plt.show()
```

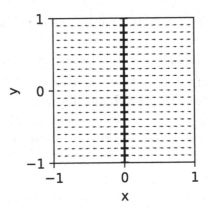

Figure 4.4

Let's examine how good this approximation is. In our code, ϵ_0 is set to 1 and $\rho = 1$. Therefore, we expect $|\vec{E}(x = D)|$ would approach 0.5. We increase the value of y_max from 1 to 2, 4, $\cdots$, 1024. As the length of the line becomes effectively infinite, compared to the length scale of analysis (between 0.1 and 1), the magnitude of the vector (vf_mag) is indeed independent of the distance D from the line.

```
# Code Block 4.6

# How good is the approximation?

def approx_line (dy,y_max):
    dx = 0.1
    x_range = np.arange(dx,1+dx,dx)
    p = np.vstack((x_range,np.zeros(len(x_range))))
    vf = np.zeros(p.shape)
    y_range = np.arange(-y_max,y_max+dy,dy)
    for ys in y_range:
        vf0, _ = get_vfield_radial_2d (p,(0,ys))
        vf = vf + (vf0)*dy
    return vf, p

dy = 0.01
y_max_range = (1,2,4,8,16,1024)
fig = plt.figure(figsize=(3,3))
```

```
for y_max in y_max_range:
    vf, p = approx_line (dy,y_max)
    vf_mag = np.sqrt(np.sum(vf**2,axis=0))
    plt.plot(p[0],vf_mag,color='gray')
    plt.text(1.05,vf_mag[-1],"%d"%y_max)

plt.xlim((0,1.3))
plt.ylim((0.2,0.6))
plt.xticks((0,0.5,1))
plt.yticks((0.3,0.4,0.5))
plt.ylabel('|E|')
plt.xlabel('x')
plt.savefig('fig_ch4_infinite_line_approx.pdf',bbox_inches='tight')
plt.show()

print('Numerical approximation = %6.5f'%(np.mean(vf_mag)))

rho = 1
print('Theoretical value = %6.5f'%(rho/2))
```

```
Numerical approximation = 0.49983
Theoretical value = 0.50000
```

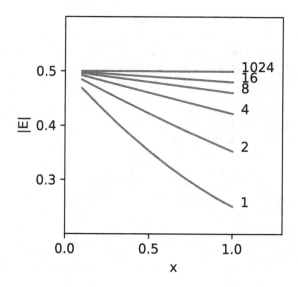

Figure 4.5

As we have seen, a linear charge distribution along an infinite line creates a vector field that points uniformly in the opposite directions on each side of the line. Thus, we can create a confined region of a uniform vector field by placing equal-and-oppositely charged lines in parallel. Both vector fields line up between the two lines, and their sum is non-zero, while the electric fields in the outer region are zero, as demonstrated by the following code block.

```python
# Code Block 4.7

# Parallel lines.

step = 0.1
x,y = np.meshgrid(np.arange(-1,1+step,step),
                  np.arange(-1,1+step,step),
                  indexing='ij')
p = np.vstack((x.flatten(),y.flatten()))
vf_double = np.zeros(p.shape)

# Choose x0 carefully,
# so that the location of the infinite line
# (namely, +x0 and -x0) does not land at
# one of the meshgrid points.
x0 = 0.55
dy = 0.1
y_max = 10
y_range = np.arange(-y_max,y_max+dy,dy)
for y0 in y_range:
    vf_r, _ = get_vfield_radial_2d (p,(+x0,y0))
    vf_l, _ = get_vfield_radial_2d (p,(-x0,y0))
    vf_double = vf_double + (vf_r - vf_l)*dy

fig = plt.figure(figsize=(2,2))
plt.scatter(np.zeros(y_range.shape)+x0,
            y_range,marker='+',color='black')
plt.scatter(np.zeros(y_range.shape)-x0,
            y_range,marker='_',color='black')
plt.quiver(p[0],p[1],vf_double[0],vf_double[1],
           angles='xy',scale_units='xy',scale=16)
plt.axis('equal')
plt.axis('square')
lim = 1.0
plt.xlim((-lim,lim))
plt.ylim((-lim,lim))
plt.xlabel('x')
plt.ylabel('y')
plt.xticks((-1,0,1))
plt.yticks((-1,0,1))
plt.savefig('fig_ch4_two_infinite_lines.pdf',bbox_inches='tight')
plt.show()
```

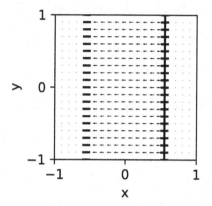

Figure 4.6

4.4   EXAMPLE: CIRCLE

Another often-studied special arrangement of electric charges is a circle. The following code block calculates electric fields at several sample locations. Outside the circle, the electric fields are radial, and their magnitudes decrease as the distance from the center increases. Inside the circle, the electric field is zero. Hence, on the sample locations inside the circle, $(-0.5, -0.5), (-0.5, 0.5), (0.5, -0.5), (0.5, 0.5)$, the electric field vectors drawn with quiver() at those locations are very short and almost invisible.

```
# Code Block 4.8

# Calculate the electric field from a circular charge distribution.
def approx_circ (p, r=1, d_phi=2*np.pi/200):
    phi_range = np.arange(0,2*np.pi,d_phi)
    vf = np.zeros(p.shape)

    for phi in phi_range:
        x0, y0 = r*np.cos(phi), r*np.sin(phi)
        vf0, _ = get_vfield_radial_2d (p,(x0,y0))
        vf = vf + (vf0)*r*d_phi
    return vf

d_phi=2*np.pi/200

scale = 1 # scale of the arrows
lim = 4.5
step = 1
```

```
x,y = np.meshgrid(np.arange(-lim,lim+step,step),
                  np.arange(-lim,lim+step,step),
                  indexing='ij')
p = np.vstack((x.flatten(),y.flatten()))
vf = approx_circ (p,d_phi=d_phi)

fig = plt.figure(figsize=(2,2))
plt.quiver(p[0],p[1],vf[0],vf[1],
           angles='xy',scale_units='xy',scale=scale)

# Draw charge distribution for illustration.
R = 1
phi_range = np.arange(0,np.pi*2,np.pi/7)
for phi in phi_range:
    x0, y0 = R*np.cos(phi), R*np.sin(phi)
    plt.scatter(x0,y0,marker='+',color='black')

plt.axis('equal')
plt.axis('square')
plt.xlim((-lim,lim))
plt.ylim((-lim,lim))
plt.xlabel('x')
plt.ylabel('y')
plt.xticks((-2,0,2))
plt.yticks((-2,0,2))
plt.savefig('fig_ch4_circle.pdf',bbox_inches='tight')
plt.show()
```

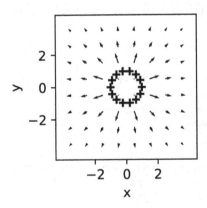

Figure 4.7

We can examine this behavior further by calculating the magnitude of the electric field along the x-axis where $y = 0$ numerically. The following plots demonstrate again that the electric field inside (shaded area) is zero, and the field outside the circle decreases. We can quantitatively verify the $\frac{1}{r}$ drop-off behavior in the outside by fitting a straight line to $\log(|\vec{E}|)$ versus $\log(r)$ pairs. The slope of the best-fit line obtained with the np.polyfit() function is -1, confirming the $\frac{1}{r}$-dependence of $|\vec{E}|$.

```
# Code Block 4.9

# Verify E is zero inside the circular shell and drops off 1/r.
px = np.array([0,0.2,0.4,0.8,1.2,1.4,1.6,1.8,2,4,8,16])
px = np.hstack((-px,px))
py = np.zeros(len(px))
p = np.vstack((px,py))

vf = approx_circ (p,d_phi=d_phi)
vf_mag = np.sqrt(np.sum(vf**2,axis=0))

fig = plt.figure(figsize=(3,3))
plt.scatter(px,vf_mag,color='black',zorder=2)

q_enc = 1
D_range = np.arange(1.2,16,0.01)
plt.plot(+D_range,q_enc/D_range,color='black')
plt.plot(-D_range,q_enc/D_range,color='black')

xbox = np.array([-1,-1,1,1])
ybox = np.array([0,q_enc/D_range[0],q_enc/D_range[0],0])
plt.fill(xbox,ybox,color='#CCCCCC',zorder=1)

plt.xlabel('D')
plt.ylabel('|E|')
plt.xticks(np.arange(-16,18,8))
plt.title('E field from a circular charge distribution')
plt.savefig('fig_ch4_circle_E.pdf',bbox_inches='tight')
plt.show()

logx, logy = np.log(px[px>1]), np.log(vf_mag[px>1])
fig = plt.figure(figsize=(3,3))
plt.scatter(logx,logy,color='black')
pfit = np.polyfit(logx,logy,1)
plt.plot(logx,logx*pfit[0]+pfit[1],color='black')
plt.xlim((0,3))
plt.xticks((0,1,2,3))
plt.xlabel('log(D)')
plt.ylabel('log(|E|)')
plt.title('log-log plot (outside)')
plt.legend((('Slope = %4.3f'%pfit[0],'log(|E|)')))
```

```
plt.savefig('fig_ch4_circle_logE.pdf',bbox_inches='tight')
plt.show()
```

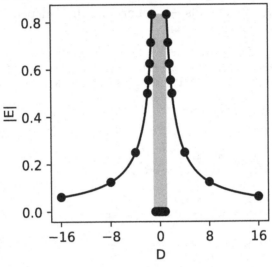

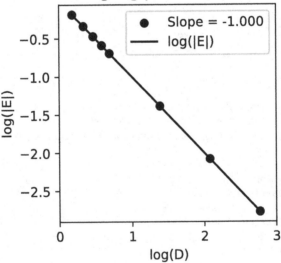

Figure 4.8

We can analytically confirm the above simulation result with some help from the sympy module. We can start with the following diagram of a circular charge distribution with radius R. The location at which we will calculate the strength of the electric field is at a distance D away from the center of the circle. The linear charge density is given by $\rho = \frac{q}{2\pi R}$, where q is the total charge on the circle.

```
# Code Block 4.10

D = 3
R = 1
step = 0.01
phi_range = np.arange(0,2*np.pi+step,step)

phi = np.pi/4
fig = plt.figure(figsize=(5,3))
plt.plot(R*np.cos(phi_range),R*np.sin(phi_range),
         color='gray',linewidth=4)
x, y = R*np.cos(phi), R*np.sin(phi)
plt.plot((0,x),(0,y),color='black')
plt.plot((x,D),(y,0),color='black')
plt.plot((0,D),(0,0),color='black')
plt.plot((x,x),(0,y),color='gray')
plt.text(x-0.5,y-0.3,'R')
plt.text(D,-0.15,'D')
plt.text(-0.1,-0.15,'0')
plt.text(x,y+0.1,'P')
plt.text(x,-0.15,'Q')
plt.text(D-1.5,+0.6,'r')
plt.text(0.2,0.07,r'$\phi$')
plt.text(D-0.7,0.07,r'$\theta$')
plt.axis('square')
plt.axis('off')
plt.xlim(np.array([-R,D])*1.1)
plt.ylim(np.array([-1,1])*R*1.1)
plt.savefig('fig_ch4_circle_diagram.pdf')
plt.show()
```

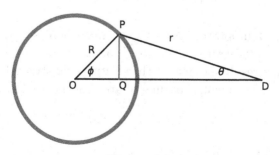

Figure 4.9

Let's consider a small infinitesimal charge element $dq = \rho R d\phi$ at point $P = (R\cos\phi, R\sin\phi)$, which is at the top of the triangle in the figure. The angle $\angle POQ$ is denoted as ϕ. The angle $\angle PDQ$ is denoted as θ. The distance r between the point of interest at D and the charge element dq is given by $r = \sqrt{(D - R\cos\phi)^2 + (R\sin\phi)^2} = \sqrt{(D^2 + R^2 - 2DR\cos\phi)}$.

The electric field due to the charge element dq at P will point along $\overline{PD}$ with the magnitude of $\frac{\rho R d\phi}{2\pi\epsilon_0 r}$ according to the two-dimensional Coulomb's law. Because there is a matching charge element across $\overline{OD}$, when the electric fields due to all charge elements around the full circle are summed, there will be no vertical component of the electric field at point D. We used the same symmetry argument to calculate the electric field of an infinite line previously. The horizontal component of electric field due to dq is then $\frac{\rho R d\phi}{2\pi\epsilon_0 r}\cos\theta$ with $\cos\theta = \frac{(D - R\cos\phi)}{r}$.

The total magnitude of $\vec{E}$ along the horizontal direction can thus be calculated by the following integral.

$$
\begin{aligned}
|\vec{E}| &= \frac{\rho}{2\pi\epsilon_0} \int_0^{2\pi} \frac{R d\phi}{r} \cos\theta \\
&= \frac{\rho}{2\pi\epsilon_0} \int_0^{2\pi} \frac{D - R\cos\phi}{D^2 + R^2 - 2DR\cos\phi} R d\phi \\
&= \frac{\rho}{2\pi\epsilon_0} \int_0^{2\pi} \frac{k - \cos\phi}{k^2 + 1 - 2k\cos\phi} d\phi \\
&= \frac{\rho}{2\pi\epsilon_0} \frac{2\pi}{k} \\
&= \frac{\rho 2\pi R}{2\pi\epsilon_0 D} \\
&= \frac{q}{2\pi\epsilon_0 D},
\end{aligned}
$$

where we have temporarily used a new dimensionless variable $k = \frac{D}{R}$. The complicated integral is equal to $\frac{2\pi}{k}$, as it can be shown numerically below. The final result shows that the electric field strength decreases as $\frac{1}{D}$, as if all charge elements, totaling q, are concentrated at the center of the circle, O.

```
# Code Block 4.11

# Numerical verification of the integral
# by summing the area under the curve.

fig = plt.figure(figsize=(6,3))

k = 2
d_phi = 0.00001
phi = np.arange(0,2*np.pi,d_phi)

i = 0
for k in (2, 0.25):
    integrand = (k-np.cos(phi))/(k**2+1-2*k*np.cos(phi))
    integral_result = np.sum(integrand)*d_phi

    print('k = %4.3f'%k)
    print('Numerical result = %8.7f'%integral_result)

    if k>1:
        print(' Expected result = %8.7f (2*pi/k for k>1)'%
            (2*np.pi/k))
    else:
        print(' Expected result = 0.0 (for 0<k<1)')
    print('')

    i = i+1
    plt.subplot(1,2,i)
    y = (k-np.cos(phi))/(k**2+1-2*k*np.cos(phi))
    plt.plot(phi,y,color='black')
    plt.fill_between(phi,y,color='gray')
    plt.title('k = %3.2f'%k)
    plt.xlabel(r'$\phi$')
    plt.ylabel(r'$\frac{k-\cos \phi}{k^2+1-2k\cos \phi}$')

plt.tight_layout()
plt.savefig('fig_ch4_numerical_integral.pdf',bbox_inches='tight')
plt.show()
```

```
k = 2.000
Numerical result = 3.1415973
 Expected result = 3.1415927 (2*pi/k for k>1)

k = 0.250
Numerical result = -0.0000063
 Expected result = 0.0 (for 0<k<1)
```

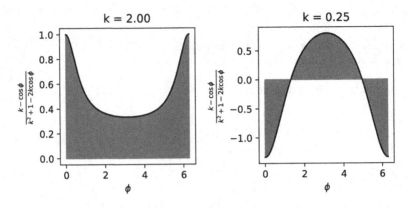

Figure 4.10

One might attempt to evaluate the above integral analytically or symbolically, but this attempt presents an interesting challenge. In the example of the circular charge distribution, the variable k implies two possibilities: the point of interest is either inside the circle($0 < k < 1$) or outside ($k > 1$). Depending on the location of the point, the distributed charges give either zero or non-zero total electric field.

The following code blocks show how the symbolic computation handles the above expression. When the value of k is specified, the symbolic integration returns the expected values of 0 or $2\pi/k$. However, when we ask sympy to perform the integral while treating k as a symbol without exactly instructing what its value is, we obtain an incorrect result of π/k instead of the correct result of $2\pi/k$.

```
# Code Block 4.12

import sympy as sym
t = sym.Symbol('t')

k = 2
print('When k = %4.3f (exterior)'%k)
print('symbolic integration returns correct result (2*pi/k for k>1):')
res = sym.integrate((k-sym.cos(t))/((k)**2+1-2*(k)*sym.cos(t)),
                    (t,0,2*sym.pi))
display(res)
print('')

k = 0.25
print('When k = %4.3f (interior)'%k)
print('symbolic integration returns correct result (0 for 0<k<1):')
```

```
res = sym.integrate((k-sym.cos(t))/((k)**2+1-2*(k)*sym.cos(t)),
                    (t,0,2*sym.pi))
print('%4.3f'%res)

print('')

# Unfortunately, sympy may not calculate
# the following integral correctly.
k = sym.Symbol('k',positive=True)

print('When k is a symbol,')
print('symbolic integration returns incorrect result:')
res = sym.integrate((k-sym.cos(t))/((k)**2+1-2*(k)*sym.cos(t)),
                    (t,0,2*sym.pi))
display(res)
```

```
When k = 2.000 (exterior)
symbolic integration returns correct result (2*pi/k for k>1):
```

π

```
When k = 0.250 (interior)
symbolic integration returns correct result (0 for 0<k<1):
0.000
```

```
When k is a symbol,
symbolic integration returns incorrect result:
```

$$\frac{\pi}{k}$$

This is an interesting lesson to remember. It is always desirable to approach a problem in multiple ways and cross-check each other, as some methods may have unexpected limitations. By the way, there is a workaround for this particular example. We can enforce the condition on the variable k with another positive variable g. If k is defined as $g+1$, we can assure that $k = g+1 > 1$. On the other hand, by asserting $k = 1-g$, we can assure that $k < 1$. These are implemented in the code blocks below, and **sympy** returns the expected answers.

```
# Code Block 4.13

g = sym.Symbol('g',positive=True)
k = sym.Symbol('k',positive=True)

k = g+1 # Ensures k>1
print('When k = g+1 > 1 with positive g,')
res = sym.integrate((k-sym.cos(t))/((k)**2+1-2*(k)*sym.cos(t)),
```

```
                            (t,0,2*sym.pi))
display(res)

print('')

k = 1-g # Ensures k<1
print('When k = 1-g < 1 with positive g,')
res = sym.integrate((k-sym.cos(t))/((k)**2+1-2*(k)*sym.cos(t)),
                      (t,0,2*sym.pi))
display(res)
```

When k = g+1 > 1 with positive g,

$$\frac{4\pi}{2g+2}$$

When k = 1-g < 1 with positive g,

$$\begin{cases} -\frac{2\pi \operatorname{sign}\left(1-\frac{2}{g}\right)}{2g-2} - \frac{2\pi}{2g-2} & \text{for } g \neq 1 \\ 0 & \text{otherwise} \end{cases}$$

4.5 GAUSS'S LAW

In the last chapter, we calculated the flux of several vector field examples and noticed an intriguing result: the flux is constant for a two-dimensional radial vector field whose magnitude decreases as $\frac{1}{r}$, regardless of the shape of a boundary. Let's pick up this topic again in this section.

We will start by considering a point charge at the origin, enclosed by differently shaped boundaries. A circular boundary that is centered at the origin is most intuitive, but the boundary is just a construct for evaluating flux, so it may be centered at a different position, shaped like an oval, or made up of several curved segments, as shown below. We may break up each boundary into tiny segments such that each piece subtends an infinitesimal angle $d\phi$ from the origin. That is, if we are looking out from the origin where the charge is located, each piece takes up a tiny portion within our field of view. The radial lines in the following figure are drawn to illustrate this, representing the decomposition of the boundary into smaller pieces. They do not represent the electric field.

```
# Code Block 4.14
```

```
step = 0.001
x = np.arange(-1,1,step)

fig = plt.figure(figsize=(6,4))

plt.subplot(1,3,1)
plt.title('(a) Circular')
y_top = np.sqrt(1-x**2)
y_bot = -y_top
plt.plot(x,y_top,color='gray',linewidth=2)
plt.plot(x,y_bot,color='gray',linewidth=2)
plt.plot([1,1],[y_top[-1],y_bot[-1]],color='gray',linewidth=2)

plt.subplot(1,3,2)
plt.title('(b) Shifted Oval')
y_top = +np.sqrt((1-x**2)*1.5)-0.4
y_bot = -np.sqrt((1-x**2)*1.5)-0.4
plt.plot(x,y_top,color='gray',linewidth=2)
plt.plot(x,y_bot,color='gray',linewidth=2)
plt.plot([1,1],[y_top[-1],y_bot[-1]],color='gray',linewidth=2)

plt.subplot(1,3,3)
plt.title('(c) Arbitrary')
y_top = 0.5*x**3+0.5
y_bot = x**2+0.5*x-0.5
plt.plot(x,y_top,color='gray',linewidth=2)
plt.plot(x,y_bot,color='gray',linewidth=2)

dtheta = np.pi/16
ray = 5
theta = np.arange(0,2*np.pi,dtheta)
zeros = np.zeros(len(theta))
for i in range(3):
    plt.subplot(1,3,i+1)
    plt.plot([zeros,ray*np.cos(theta)],
             [zeros,ray*np.sin(theta)],
             color='gray',linewidth=0.5)
    plt.scatter(0,0,marker='+',color='black')
    plt.axis('square')
    plt.axis('off')
    plt.xlim(np.array([-1,1])*2)
    plt.ylim(np.array([-1,1])*2)

plt.tight_layout()
plt.savefig('fig_ch4_gauss_law_boundary.pdf',bbox_inches='tight')
plt.show()
```

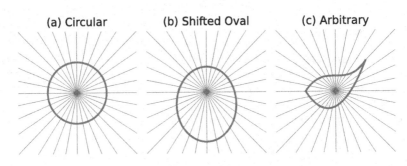

Figure 4.11

We can prove that the flux of the electric field is independent of the radius of the enclosing boundary, if this two-dimensional vector field has $\frac{1}{r}$-dependence according to Coulomb's Law. Because the spatial extent of a boundary also has r-dependence, these factors cancel each other out. In other words, consider a point charge $+q$ and a circular boundary centered at the location of the point. The normal and the electric field vectors are parallel to each other, so that $\vec{E}\cdot\hat{n} = |\vec{E}| = \frac{q}{2\pi\epsilon_0 r}$. Furthermore, differential length dl is given by $rd\phi$, where ϕ is the angle around the location of the charge. Then, the flux through the piece of boundary that subtends $d\phi$ is

$$d\Phi_{\mathbf{E}} = \vec{E} \cdot \hat{n}dl = |\vec{E}|dl = \frac{q}{2\pi\epsilon_0 r}rd\phi = \frac{qd\phi}{2\pi\epsilon_0},$$

which is independent of r. For larger r (or for a piece of flux boundary farther away from the source charge), the electric field is weaker by a factor of $\frac{1}{r}$, and the extent of the boundary is again larger by a factor of r ($dl = rd\phi$), as long as the boundary segments cover the same angular extent $d\phi$. The electric field is stronger for smaller r, but the extent of the boundary through which the electric field penetrates is smaller. These two variations in r cancel out. In three dimensions where the surface area of the boundary goes like r^2 (or $da = r^2 \sin\theta\, d\theta\, d\phi$ in spherical coordinate), the electric field that has $\frac{1}{r^2}$-dependence would have the same property.

```
# Code Block 4.15

lw = 6
scale = 0.2
dphi = 0.2
loc = [5,9]
```

```
fig = plt.figure(figsize=(6,4))
gs = fig.add_gridspec(2,1)

ax_titles = ['(a) near','(b) far-away']
for i in range(2):
    ax = fig.add_subplot(gs[i])
    ax.set_title(ax_titles[i])
    ax.scatter(0,0,marker='+',color='black')
    ax.plot([0,10],[0,+dphi],color='gray',linewidth=1)
    ax.plot([0,10],[0,-dphi],color='gray',linewidth=1)

    ax.plot(np.array([1,1])*loc[i],np.array([-1,1])*dphi*loc[i]/10,
            color='black',linewidth=lw,alpha=0.25)
    x,y = np.meshgrid(loc[i],np.arange(-0.3,0.4,0.15),indexing='ij')

    ax.quiver(x,y,x/(x**2+y**2),y/(x**2+y**2),
              angles='xy',scale_units='xy',scale=scale,
              color='gray',linewidth=1)
    ax.set_xlim((-1,11))
    ax.set_ylim((-0.5,0.5))
    ax.axis('off')

plt.tight_layout()
plt.savefig('fig_ch4_gauss_boundary_distance.pdf',bbox_inches='tight')
plt.show()
```

(a) near

(b) far-away

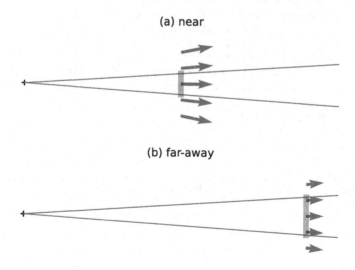

Figure 4.12

The independence of flux is not only for r but also for the orientation of the boundary. In the earlier chapter, we compared the flux through a perpendicular and slanted boundary for a uniform vector field. Now, let's extend this idea for a $\frac{1}{r}$-vector field in two dimensions.

```python
# Code Block 4.16

lw = 6
scale = 0.2
dphi = 0.2
loc = 7

fig = plt.figure(figsize=(6,4))
gs = fig.add_gridspec(2,1)

ax_titles = ['(a) Perpendicular','(b) slanted']
for i in range(2):
    ax = fig.add_subplot(gs[i])
    ax.set_title(ax_titles[i])
    ax.scatter(0,0,marker='+',color='black')
    ax.plot([0,10],[0,+dphi],color='gray',linewidth=1)
    ax.plot([0,10],[0,-dphi],color='gray',linewidth=1)

    xs = i*0.7 # determines the amount of slant
    ax.plot(np.array([loc+xs,loc-xs]),np.array([-1,1])*dphi*loc/10,
            color='black',linewidth=lw,alpha=0.5)
    x,y = np.meshgrid(np.array([loc-0.8,loc,loc+0.8]),
                      np.arange(-0.3,0.4,0.15),
                      indexing='ij')
    ax.quiver(x,y,x/(x**2+y**2),y/(x**2+y**2),
              angles='xy',scale_units='xy',scale=scale,
              color='gray',linewidth=1)

    ax.set_xlim((-1,11))
    ax.set_ylim((-0.5,0.5))
    ax.axis('off')

plt.tight_layout()
plt.savefig('fig_ch4_gauss_boundary_orientation.pdf')
plt.show()
```

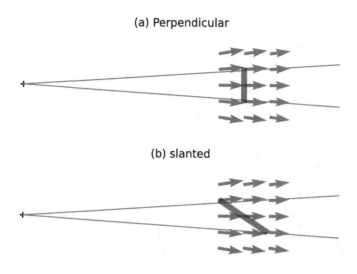

(a) Perpendicular

(b) slanted

Figure 4.13

We can denote the relative angle between the perpendicular and slanted boundaries as θ, as shown in the following figure. Then, the relative angle between their respective normal vectors is also θ. In the case of a perpendicular boundary, the dot product between the vector field $\vec{v}$ and the normal vector $\hat{n}_\perp$ of the perpendicular boundary is $\vec{v} \cdot \hat{n}_\perp = |\vec{v}|$. In the case of a slanted boundary, the dot product between $\vec{v}$ and the normal vector of the slanted boundary is $\vec{v} \cdot \hat{n}_s = |\vec{v}| \cos \theta$. At the same time, the length of the slanted boundary is bigger by a factor of $\frac{1}{\cos \theta}$. In other words, $dl_s = \frac{dl_\perp}{\cos \theta}$.

Therefore, the amount of flux through a differential extent $dl_\perp$ of a perpendicular boundary is $d\Phi_\perp = |\vec{v}|dl_\perp$, and the flux through the slanted boundary is $d\Phi_s = |\vec{v}| \cos \theta dl_s = |\vec{v}| \cos \theta \frac{dl_\perp}{\cos \theta}$, which shows that $d\Phi_\perp = d\Phi_s$. Hence, the orientation of the boundary does not change the flux. Note we were considering an infinitesimal segment of a boundary, such that the two rays from the source charge $d\phi$ apart are considered parallel in the limit as $d\phi$ goes to zero.

```
# Code Block 4.17

lw = 12
fs = 14
fig = plt.figure(figsize=(4,4))
```

```
# Perpendicular boundary
plt.plot([+0.0,-0.0],[-0.5,+0.5],color='black',linewidth=lw)
plt.quiver(0,0,1,0,
           angles='xy',scale_units='xy',scale=2,color='black')
plt.text(-0.15,0.35,r'$dl_{\perp}$',fontsize=fs)
plt.text(0.5,0.05,r'$\hat{n}_{\perp}$',fontsize=fs)

# Slanted boundary
plt.plot([+0.5,-0.5],[-0.5,+0.5],color='gray',linewidth=lw)
plt.quiver(0,0,0.7,0.7,
           angles='xy',scale_units='xy',scale=2,color='gray')
plt.text(-0.4,0.2,r'$dl_{s}$',fontsize=fs)
plt.text(0.3,0.4,r'$\hat{n}_{s}$',fontsize=fs)

plt.plot([0,-0.5],[+0.5,+0.5],color='gray',linewidth=1,ls='--')
plt.plot([0,+0.5],[-0.5,-0.5],color='gray',linewidth=1,ls='--')
plt.text(-0.1,0.15,r'$\theta$',fontsize=fs)
plt.text(0.15,0.05,r'$\theta$',fontsize=fs)

plt.axis('square')
plt.axis('off')
plt.tight_layout()
plt.savefig('fig_ch4_perp_vs_slanted.pdf',bbox_inches='tight')
plt.show()
```

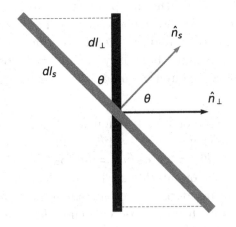

Figure 4.14

We have just shown that for a special vector field that has the $\frac{1}{r}$-dependence, the flux is independent of r and θ (the distance from a source charge and the orientation of the boundary). From this independence, it follows that a circular boundary enclosing a point charge and a differently shaped enclosure surrounding the same charge will pass the same amount of flux, because any boundary can be broken up into infinitesimal pieces that subtend the same angular extents as the pieces from a circular boundary do.

The total flux, the sum of flux through each boundary segment, can be expressed as an integral: $\Phi_E = \oint \vec{E} \cdot \hat{n} dl$, where $\oint$ indicates an integral over a closed boundary.

When this boundary is a circle with radius r centered at the location of a point charge q, the total flux can be calculated easily.

$$\Phi_E = \oint \vec{E} \cdot \hat{n} dl = \int_0^{2\pi} \frac{q}{2\pi\epsilon_0 r} r d\phi = \frac{q}{\epsilon_0}.$$

The above value will be valid for any arbitrarily shaped boundary that encloses the same q.

If there are multiple point charges (q_1, q_2, $\cdots$) or distribution of charge over a finite volume, we can use the superposition principle and perform the flux calculation over a boundary that encloses these charges.

$$
\begin{aligned}
\Phi_E = \oint \vec{E} \cdot \hat{n} dl &= \oint (\vec{E}_1 + \vec{E}_2 + \cdots) \cdot \hat{n} dl \\
&= \oint \vec{E}_1 \cdot \hat{n} dl + \oint \vec{E}_2 \cdot \hat{n} dl + \cdots \\
&= \frac{q_1 + q_2 + \cdots}{\epsilon_0} \\
&= \frac{q_{enc}}{\epsilon_0},
\end{aligned}
$$

where q_{enc} is the total amount of charge enclosed by the boundary.

This remarkable conclusion is known as Gauss's law. The enclosing boundary is known as a Gaussian surface. The term "surface" comes from the fact that in three dimensions, a boundary is a two-dimensional surface, just like the flexible membrane of a balloon that encloses a volume of air. In two dimensions, a boundary is a one-dimensional closed contour.

4.6 NUMERICAL VERIFICATION OF GAUSS'S LAW

Let us verify Gauss's law numerically for a few cases. The following series of code blocks set up single or multiple electric charges in space and calculate the electric flux through different boundary configurations. For each case, we can see that the pattern of the electric field on the boundary may be complicated and asymmetric, but the total flux is always equal to the number of enclosed charges divided by ϵ_0, which is set to 1 in the code. This result holds whether the boundary is a circle, a square, or any other shape.

```python
# Code Block 4.18

# The following functions have already been defined
# and used in the earlier chapters.

# Excerpt from Code Block 3.14

def get_normals_enc (boundary,inside=(0,0)):
    boundary_ext = np.hstack((boundary,boundary[:,:2]))
    x, y = boundary_ext[0], boundary_ext[1]

    very_small_num = 10**(-10) # avoid divide by zero.
    slope = (y[2:]-y[:-2])/(x[2:]-x[:-2] + very_small_num)
    norm_vec_slope = -1/(slope + very_small_num)
    u, v = 1, norm_vec_slope
    mag = np.sqrt(u**2+v**2)
    u, v = u/mag, v/mag
    n = np.vstack((u,v))

    # Calculate the sign of a dot product between n and
    # vector between (x,y)-inside, which determines
    # which way n should point.
    x, y = x[1:-1]-inside[0], y[1:-1]-inside[1]
    dot_prod_sign = np.sign(x*n[0] + y*n[1])
    n = n*dot_prod_sign
    return n

def plot_normals_enc (boundary,ax,inside=(0,0),scale=3):
    n = get_normals_enc(boundary,inside=inside)

    boundary_ext = np.hstack((boundary,boundary[:,:2]))
    x, y = boundary_ext[0], boundary_ext[1]

    color = '#CCCCCC'
    ax.scatter(x,y,color='gray')
    ax.plot(x,y,color='gray')
    # Mark the inside point with a diamond marker.
    #ax.scatter(inside[0],inside[1],marker='d',color='black')
    ax.quiver(x[1:-1],y[1:-1],n[0],n[1],color=color,
              angles='xy',scale_units='xy',scale=scale)
    ax.set_xlabel('x')
```

```
    ax.set_ylabel('y')
    ax.axis('equal')
    ax.axis('square')
    return

# Excerpt from Code Block 3.15

def points_along_circle (r=1, step=np.pi/10):
    phi = np.arange(-np.pi,np.pi,step)
    p = np.vstack((r*np.cos(phi),r*np.sin(phi)))
    return p

def points_along_square (s=1, step=0.1):
    # Get points along a square, whose sides are of length s.
    d = s/2
    one_side = np.arange(-d,d,step)
    N = len(one_side)
    p_top = np.vstack((+one_side,np.zeros(N)+d)) # top side
    p_rgt = np.vstack((np.zeros(N)+d,-one_side)) # right side
    p_bot = np.vstack((-one_side,np.zeros(N)-d)) # bottom side
    p_lft = np.vstack((np.zeros(N)-d,+one_side)) # left side
    p = np.hstack((p_top,p_rgt,p_bot,p_lft))
    return p

# Excerpt from Code Block 3.17

def get_flux_enc (boundary,vfield,inside=(0,0)):
    boundary_ext = np.hstack((boundary,boundary[:,:2]))
    x, y = boundary_ext[0], boundary_ext[1]
    dl_neighbor = np.sqrt((x[1:]-x[:-1])**2+(y[1:]-y[:-1])**2)
    dl = (0.5)*(dl_neighbor[1:]+dl_neighbor[:-1])
    n = get_normals_enc(boundary,inside=inside)
    xv, yv = vfield[0], vfield[1]
    dotprod = xv*n[0] + yv*n[1]
    flux = np.sum(dl*dotprod)
    return flux
```

```
# Code Block 4.19

# Verifying Gauss's law numerically.

def verify_gauss (q,p_charges,boundary_type='circle'):

    # q: sign of each point charge (either +1 or -1)
    # p_charges: location of point charge

    # Use small step to make the calculation more precise.
    # Use large step to make plots less cluttered with arrows.
    if boundary_type=='circle':
```

```
        p = points_along_circle(r=1,step=0.001)
        p_plot = points_along_circle(r=1,step=0.3)
    else:
        p = points_along_square(s=2,step=0.001)
        p_plot = points_along_square(s=2,step=0.3)

    total_vf = np.zeros(p.shape)
    total_vf_plot = np.zeros(p_plot.shape)

    # Loop through each point charge.
    for i, p_charge in enumerate(p_charges):
        vf, _ = get_vfield_radial_2d (p,p_charge)
        total_vf = total_vf + vf*q[i]

        vf, _ = get_vfield_radial_2d (p_plot,p_charge)
        total_vf_plot = total_vf_plot + vf*q[i]

    flux = get_flux_enc(p,total_vf)

    plt.figure(figsize=(3,3))
    for i in range(len(q)):
        if q[i] > 0:
            marker = '+'
        else:
            marker = '_'
        plt.scatter(p_charges[i,0],p_charges[i,1],
                    marker=marker,s=100,color='black')

    plot_normals_enc(p_plot,plt.gca(),inside=(0,0))
    plt.quiver(p_plot[0],p_plot[1],
               total_vf_plot[0],total_vf_plot[1],
               color='black',scale=2)
    plt.xlim((-2,2))
    plt.ylim((-2,2))
    plt.xticks((-1,0,1))
    plt.yticks((-1,0,1))
    plt.title('Flux = %4.3f'%(flux))
    return

# Create and examine many examples.

verify_gauss ([1],np.array([[0,0]]))
plt.savefig('fig_ch4_verify_gauss_example01.pdf')
plt.show()

verify_gauss ([1],np.array([[0.4,0.4]]))
plt.savefig('fig_ch4_verify_gauss_example02.pdf')
plt.show()
```

```
verify_gauss ([1],np.array([[1.1,1.1]]))
plt.savefig('fig_ch4_verify_gauss_example03.pdf')
plt.show()

verify_gauss ([1],np.array([[0,0]]),boundary_type='square')
plt.savefig('fig_ch4_verify_gauss_example04.pdf')
plt.show()

verify_gauss ([1],np.array([[0.4,0.4]]),boundary_type='square')
plt.savefig('fig_ch4_verify_gauss_example05.pdf')
plt.show()

verify_gauss ([1],np.array([[0.7,1.4]]),boundary_type='square')
plt.savefig('fig_ch4_verify_gauss_example06.pdf')
plt.show()

verify_gauss ([1,-1],np.array([[-0.4,-0.4],[0.3,0.3]]))
plt.savefig('fig_ch4_verify_gauss_example07.pdf')
plt.show()

verify_gauss ([1,-1,1],np.array([[0,0],[0.3,0.3],[-0.2,0.1]]))
plt.savefig('fig_ch4_verify_gauss_example08.pdf')
plt.show()

verify_gauss ([1,-1,-1,1],
             np.array([[0,0],[0.3,0.3],[-0.4,0.1],[-0.1,-0.4]]))
plt.savefig('fig_ch4_verify_gauss_example09.pdf')
plt.show()

verify_gauss ([1,1,-1,1],
             np.array([[0,0],[-0.2,0.2],[-0.4,0.1],[1.4,1.4]]))
plt.savefig('fig_ch4_verify_gauss_example10.pdf')
plt.show()

verify_gauss ([1,-1,-1,1],
             np.array([[0,0],[0.3,0.3],[-0.4,0.1],[-0.1,-0.4]]),
             boundary_type='square')
plt.savefig('fig_ch4_verify_gauss_example11.pdf')
plt.show()

verify_gauss ([1,1,-1,1],
             np.array([[0,0],[-0.2,0.2],[-0.4,0.1],[1.4,1.4]]),
             boundary_type='square')
plt.savefig('fig_ch4_verify_gauss_example12.pdf')
plt.show()
```

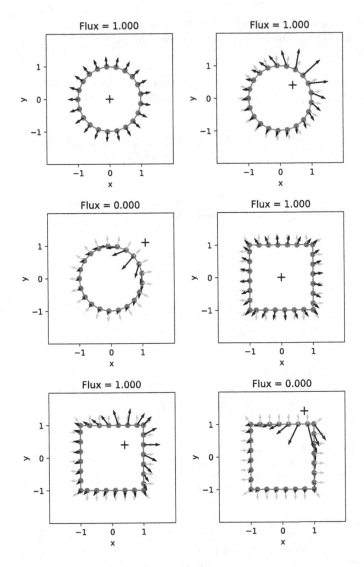

Figure 4.15

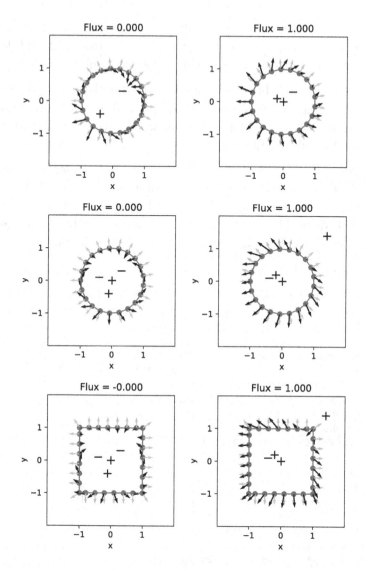

Figure 4.16

4.7   LINE AND CIRCLE, AGAIN

As the previous examples illustrate, the electric field from a charge distribution can be quite complicated and difficult to be studied analytically. However, when there is an exploitable symmetry in the charge distribution, Gauss's law can be used to calculate the electric field efficiently. Such special cases include charges distributed along an infinite line or a circle, which we have already analyzed using Coulomb's Law and the superposition principle. Here, we return to these examples and demonstrate how Gauss's law quickly yields the same results as before.

For a vertical infinite line with a uniform charge density ρ, the electric field will only have an $\hat{x}$-component. $\vec{E}(x, y)$ will point toward either $+\hat{x}$ or $-\hat{x}$, as we discussed before. In other words, the electric field can be written as $\vec{E}(x, y) = \pm|\vec{E}(x)|\hat{x}$.

Given this symmetry, we can consider a square boundary, so that the dot products between the electric field and normal vectors of the boundary would be simple scalar values of either zero or the magnitude of the electric field. The plot below shows the infinite line and a square boundary with a side of $2D$, enclosing a section of the line.

```
# Code Block 4.20

fig = plt.figure(figsize=(3,3))
# Infinite line
D = 0.5
step = 0.1
x,y = np.meshgrid(np.arange(-1,1+step,step),
                  np.arange(-1,1+step,step),
                  indexing='ij')
p = np.vstack((x.flatten(),y.flatten()))
# Avoid choosing positions along the y-axis.
idx = np.where(np.abs(p[0])>0.01)
p = np.vstack((p[0,idx],p[1,idx]))

dy = 0.1
y_max = 10
y_range = np.arange(-y_max,y_max,dy) # Practically infinite.

vf_total = np.zeros(p.shape)
for ys in y_range:
    vf, p = get_vfield_radial_2d (p,(0,ys))
    vf_total = vf_total + (vf)*dy

plt.plot([-D,D,D,-D,-D],[D,D,-D,-D,D],color='gray',
         linewidth=8,alpha=0.5)

p_square = points_along_square(s=D*2,step=0.2)
plot_normals_enc(p_square,plt.gca())

plt.scatter(np.zeros(y_range.shape),y_range,
            marker='+',color='black')
plt.quiver(p[0],p[1],vf_total[0],vf_total[1],color='black',
```

```
                angles='xy',scale_units='xy',scale=8)
plt.axis('equal')
plt.axis('square')
plt.xlabel('x')
plt.ylabel('y')
lim = 1.25
plt.xlim((-lim,lim))
plt.ylim((-lim,lim))
plt.xticks((-1,0,1))
plt.yticks((-1,0,1))
plt.savefig('fig_ch4_infinite_line_gauss.pdf',bbox_inches='tight')
plt.show()
```

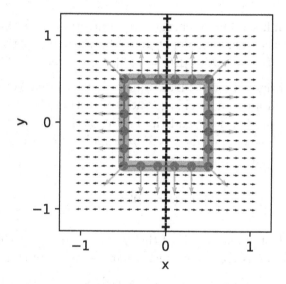

Figure 4.17

We can easily calculate the flux through the square's four sides and add them. This is the left-hand side of Gauss's law. The flux is zero for the top and bottom sides because the electric field and the normal vectors are perpendicular to each other, making their dot products equal to zero.

For the boundary on the left at $x = -D$, $\vec{E}(x = -D, y) \cdot \hat{n} = (-|\vec{E}(x = -D)|\hat{x}) \cdot (-\hat{x}) = |\vec{E}(x = -D)|$. Likewise at the boundary on the right at $x = +D$, $\vec{E}(x = +D, y) \cdot \hat{n} = (+|\vec{E}(x = +D)|\hat{x}) \cdot (+\hat{x}) = |\vec{E}(x = +D)|$. The total flux, as we go around the square in a counterclockwise direction, becomes

$$
\begin{aligned}
\Phi_{\mathbf{E}} &= \oint \vec{\mathbf{E}} \cdot \hat{n} dl \\
&= \int_{\text{top}} \vec{\mathbf{E}} \cdot \hat{n} dl + \int_{\text{left}} \vec{\mathbf{E}} \cdot \hat{n} dl + \int_{\text{bottom}} \vec{\mathbf{E}} \cdot \hat{n} dl + \int_{\text{right}} \vec{\mathbf{E}} \cdot \hat{n} dl \\
&= 0 + \int_{+D}^{-D} \vec{\mathbf{E}}(x = -D) \cdot \hat{n}(-dy) + 0 + \int_{-D}^{+D} \vec{\mathbf{E}}(x = +D) \cdot \hat{n} dy \\
&= 2D \left(|\vec{\mathbf{E}}(x = -D)| + |\vec{\mathbf{E}}(x = +D)| \right).
\end{aligned}
$$

Because this infinite line of charge has symmetry about the y-axis, the magnitude of the field at $x = +D$ would be the same as the magnitude at $x = -D$, or $|\vec{\mathbf{E}}(x = -D)| = |\vec{\mathbf{E}}(x = +D)|$. Hence, $\Phi_{\mathbf{E}} = 4D|\vec{\mathbf{E}}(x = D)|$.

Now, let's consider the right-hand side of Gauss's law. We need to calculate the total amount of charge enclosed within the boundary. With the linear charge density ρ and the height of the square being $2D$, the net charge within the boundary is $2D\rho$. Finally, when we equate the two sides of Gauss's law, we arrive at

$$
4D|\vec{\mathbf{E}}(x = D)| = \frac{2D\rho}{\epsilon_0}.
$$

In other words, the magnitude of the electric field is $\frac{\rho}{2\epsilon_0}$, which does not depend on D and is consistent with what we had obtained previously.

The second example is a ring with a radius of R. As before, we will assume that the total charge q is uniformly distributed so that the linear charge density of the ring is $\rho = \frac{q}{2\pi R}$. Given the circular symmetry, we expect the electric field of this charge distribution to be radial, $\hat{r}$. Next, suppose we enclose this charged ring with a circular boundary of radius D, as shown below. The normal vectors of a circular boundary are radial, too, so $\hat{n} = \hat{r}$.

Hence, the dot product between $\vec{\mathbf{E}}$ due to the ring and the normal vector $\hat{n}$ of the circular boundary is $\vec{\mathbf{E}} \cdot \hat{n} = |\vec{\mathbf{E}}|\hat{r} \cdot \hat{r} = |\vec{\mathbf{E}}|$. The flux through the circular boundary is $\Phi_{\mathbf{E}} = \oint \vec{\mathbf{E}} \cdot \hat{n} dl = \int_0^{2\pi} |\vec{\mathbf{E}}| D d\phi = 2\pi D|\vec{\mathbf{E}}|$.

```
# Code Block 4.21

# Circular charge distribution with radius = 1.

scale = 0.8 # scale of the arrows
step = 1
lim = 4.5
x, y = np.meshgrid(np.arange(-lim,lim+step,step),
                   np.arange(-lim,lim+step,step),
                   indexing='ij')
p = np.vstack((x.flatten(),y.flatten()))
vf = approx_circ(p)

fig = plt.figure(figsize=(3,3))
plt.quiver(p[0],p[1],vf[0],vf[1],
           angles='xy',scale_units='xy',scale=scale)

p_circle = points_along_circle(r=2.5,step=0.25)
plt.plot(p_circle[0],p_circle[1],color='gray',
         linewidth=8,alpha=0.5)
plot_normals_enc(p_circle,plt.gca())

# Draw charge distribution for illustration.
for phi in np.arange(0,np.pi*2,np.pi/7):
    x0, y0 = np.cos(phi), np.sin(phi)
    plt.scatter(x0,y0,marker='+',color='black')

plt.axis('equal')
plt.axis('square')
plt.xlim((-lim,lim))
plt.ylim((-lim,lim))
plt.xlabel('x')
plt.ylabel('y')
plt.xticks((-2,0,2))
plt.yticks((-2,0,2))
plt.text(+2.9,0,'D')
plt.text(+1.4,0,'R')
plt.savefig('fig_ch4_circle_gauss.pdf',bbox_inches='tight')
plt.show()
```

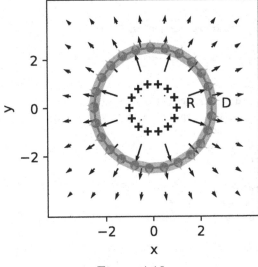

Figure 4.18

The next step is to consider the right-hand side of Gauss's law and calculate the net charge inside the boundary. For $D > R$, all the charges are enclosed by the boundary, so $q_{enc} = q$. For $D < R$, where the boundary is inside of the charged ring, $q_{enc} = 0$. Since $\Phi_{\mathbf{E}} = 2\pi D|\vec{\mathbf{E}}| = q_{enc}/\epsilon_0$, $|\vec{\mathbf{E}}| = 0$ inside the ring. Outside the ring, $|\vec{\mathbf{E}}| = \frac{q}{2\pi\epsilon_0 D}$, so the electric field drops inversely as D increases. These results are consistent with our earlier simulation results.

These two special examples with an infinite line and a ring demonstrate that an electric field can be easily determined from Gauss's law when there is an exploitable symmetry in the geometry of the charge distribution.

4.8 ELECTRIC FIELD IN THREE-DIMENSION

All of the previous discussions about electricity(Coulomb's law, Gauss's law, etc.) have been presented as if we live in a two-dimensional space for ease of visual illustration and mathematical simplicity. However, we live in a three-dimensional space, so several equations must be updated.

It is convenient to consider a displacement vector, which is the "line of sight" between the source point (x_s, y_s, z_s) and the point of interest (x, y, z) at which the vector field is being calculated. The displacement vector $\vec{r} = (x - x_s)\hat{x} + (y - y_s)\hat{y} + (z - z_s)\hat{z}$, and $r = \sqrt{(x - x_s)^2 + (y - y_s)^2 + (z - z_s)^2}$. The electric field from a point charge acts along $\vec{r}$, and in three dimensions, the electric field strength decreases inversely with the squared distance, so the three-dimensional Coulomb's law is expressed as the following:

$$\vec{E}_{3D}(x, y, z) = \frac{q}{4\pi\epsilon_0} \frac{\vec{r}}{r^3} = \frac{q}{4\pi\epsilon_0} \left(\frac{x - x_s}{r^3}\hat{x} + \frac{y - y_s}{r^3}\hat{y} + \frac{z - z_s}{r^3}\hat{z} \right).$$

The expression $\frac{\vec{r}}{r^3}$ signifies that the electric field is directed radially from the source charge, and the field strength decreases as $\frac{1}{r^2}$.

Then,

$$|\vec{E}_{3D}| = \frac{q}{4\pi\epsilon_0} \frac{1}{r^2}.$$

The formulation of Gauss's law remains the same:

$$\oint \vec{E} \cdot \hat{n} \, da = \frac{q_{enc}}{\epsilon_0},$$

where the three-dimensional volume that encloses q_{enc} has a two-dimensional boundary surface.

We already discussed the case of $\vec{E}_{2D}$, which has $\frac{1}{r}$-dependence in two dimensions. In three dimensions, $\vec{E}_{3D}$ varies with $\frac{1}{r^2}$-dependence. As we integrate the three-dimensional electric fields over a two-dimensional boundary surface whose spatial extent increases with r^2, the total flux is a constant value that is solely determined by the amount of the enclosed charge.

4.9 DIVERGENCE

The divergence of a vector field $\vec{K}$ is defined as

$$\nabla \cdot \vec{K} = \frac{\partial K_x}{\partial x} + \frac{\partial K_y}{\partial y} + \frac{\partial K_z}{\partial z}.$$

The divergence is a scalar quantity that represents the density of the outward flux of a vector field. For example, let's consider a simple uniform vector field, $\vec{K}(x, y, z) = \hat{x}$, as shown in Figure 3.1. Its divergence is zero. Pick any confined region in the space, and we will see that the vector field flows in from the left side and flows out to the right side along the x-axis. There is no source of the flow within the region, and the total flux equals zero.

Now consider a different vector field, $\vec{K} = x\hat{x}$, whose divergence is $+1$. In this case, there is more flux leaving the confined region than entering. In other words, there is a source of the vector field inside the region. With a vector field $\vec{K} = -x\hat{x}$, the divergence is -1, indicating that there is more flux entering than leaving. In other words, there is a sink of the vector field.

It is also instructive to consider a vector field of the form, $\vec{K} = f(y,z)\hat{x}$, where the function $f(y,z)$ does not depend on x. No matter how complex this function is, $\frac{\partial f(y,z)}{\partial x} = 0$, so the divergence of $\vec{K}$ is zero. The amount of flux entering and leaving through a surface perpendicular to the x-axis will always be equal to each other. In other words, there is no source or sink for this vector field inside any confined region of space.

The divergence theorem captures this idea:

$$\iiint_{\text{Volume}} (\nabla \cdot \vec{K}) dV = \oiint_{\text{Area}} \vec{K} \cdot d\vec{a},$$

where the volume for the triple integral on the left-hand side is defined by the enclosing surface for the double integral on the right-hand side.

4.10 GAUSS'S LAW, AGAIN

Using the divergence theorem, let us examine Gauss's law again. We can replace $\vec{K}$ with $\vec{E}$, so that

$$\oiint_{\text{Area}} \vec{E} \cdot d\vec{a} = \iiint_{\text{Volume}} (\nabla \cdot \vec{E}) dV.$$

According to Gauss's law, for a given charge density ρ

$$\oiint_{\text{Area}} \vec{E} \cdot d\vec{a} = \frac{q_{\text{enc}}}{\epsilon_0} = \frac{1}{\epsilon_0} \iiint_{\text{Volume}} \rho dV.$$

The integrands of the right-hand sides of these two expressions must be equal to each other since these equalities must hold for any arbitrarily shaped boundary and the volume it encloses. In other words,

$$\nabla \cdot \vec{E} = \frac{1}{\epsilon_0}\rho.$$

This is the differential form of Gauss's law.

## CHAPTER 5

Magnetic Field

5.1 ELECTRIC CURRENTS AND BIOT-SAVART LAW

A continuous flow of electric charges is called an electric current, and it generates another special vector field known as a magnetic field. A Danish physicist Oersted is known to have first noticed the deflection of a compass needle around a current-carrying wire. Subsequent experimental and theoretical studies have led to the following mathematical expression for magnetic vector field $\vec{\mathbf{B}}$, due to an arbitrarily shaped wire with a steady current of magnitude I.

$$\vec{\mathbf{B}}(x,y,z) = \frac{\mu_0}{4\pi} \int_{\text{along wire}} \frac{I d\vec{\mathbf{l}} \times \vec{\mathbf{r}}}{r^3}.$$

The above expression, known as the Biot-Savart law, states that the direction and magnitude of a magnetic field at position (x, y, z) can be calculated by integrating all magnetic fields generated by infinitesimal sections of steady electric currents at position (x_s, y_s, z_s) on a wire. As we have discussed in the context of an electric field, the displacement vector is defined as: $\vec{\mathbf{r}} = (x - x_s)\hat{\mathbf{x}} + (y - y_s)\hat{\mathbf{y}} + (z - z_s)\hat{\mathbf{z}}$, and its magnitude is: $r = \sqrt{(x - x_s)^2 + (y - y_s)^2 + (z - z_s)^2}$.

This is the same superposition principle we have already encountered in the context of calculating the electric field: when there is a static distribution of electric charges (for example, along a line or along a circle), the total electric field is the sum of the electric fields due to individual charge elements. For magnetic fields, instead of an individual electrical charge, the source of the magnetic field is the movement of electric charges or individual current element $I d\vec{\mathbf{l}}$.

 DOI: 10.1201/9781003397496-5

Similar to Coulomb's law in three dimensions, the Biot-Savart law describes that the strength of the magnetic field due to the infinitesimal current at point (x_s, y_s, z_s) in a wire decreases as $\frac{1}{r^2}$. $\vec{dl}$ is an infinitesimal length element along the wire. If I has a positive sign, the current will flow in the same direction as in $\vec{dl}$, and if I is negative, its direction is opposite to that of $\vec{dl}$. A very interesting and unintuitive fact is that the direction of the magnetic field is determined by the cross product between the direction of the current $I\vec{dl}$ and the displacement vector $\vec{r}$.

The proportionality constant μ_0 is known as the permeability of free space with a value of $4\pi \times 10^{-7}$ T m A^{-1}. When the current and distance are expressed in the standard units of ampere (A) and meter (m), this constant returns the magnetic field in a unit of tesla (T). We will consider several examples of magnetic fields generated by different patterns of current flow.

This figure illustrates the relationship between (x, y, z), (x_s, y_s, z_s), the displacement vector $\vec{r}$, and the infinitesimal length segment $\vec{dl}$.

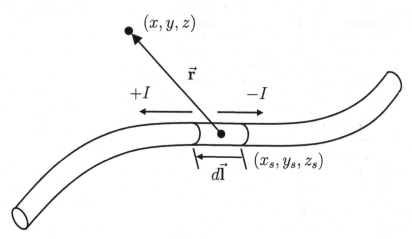

Figure 5.1

As a starting example, consider a straight wire with length L that carries a steady current I. Steady current means that the number of charges flowing into any point in a wire equals the number of charges exiting the point at any given time, so there is no net accumulation of electric charges. Hence, in this situation of a wire with finite length, we will assume that the electric charges, upon reaching one end, somehow loop back to the other end or that the wire stretches to infinity. For mathematical simplicity, we will further assume that the straight wire passes through the origin and stretches along the z-axis, and the current flows in the $+\hat{z}$ direction. The current element $Id\vec{l}$ in the Biot-Savart law is then $+Idz\hat{z}$.

The following code block defines and plots this situation. The function `currents_along_line()` takes multiple input arguments for defining the segment length dL, the total length L of the wire, and the magnitude of steady current I. The outputs of the function are two arrays that specify $Id\vec{l}$ at various points along the wire. The first output array p specifies (x_s, y_s, z_s), and the second output array curr contains the current segments $Id\vec{l}$ at those positions. In the case of a straight wire along the z-axis, the current segments only have the z component of magnitude I. We use the `quiver()` function to display the current segments.

```python
# Code Block 5.1

import numpy as np
import matplotlib.pyplot as plt

def currents_along_line (I,L,dL,x0=0,y0=0):
    # Current distribution for a straight wire,
    # positioned vertically at (x0,y0).
    # I: total current through the wire.
    # L: length of wire along z.
    # dL: infinitesimal segment of wire (L >> dL)
    # p = (x0,y0,z)
    # curr = (0,0,I*dL)
    vec = np.arange(-L/2,L/2+dL,dL)
    N = len(vec)
    p = np.zeros((3,N))
    p[0], p[1] = x0, y0
    p[2] = vec

    curr = np.zeros((3,N))
    curr[2] = I*dL
    return p, curr

def tidy_ax(ax):
    ax.set_xlim((-1,1))
    ax.set_ylim((-1,1))
    ax.set_zlim((-1,1))
    ax.set_xticks((-1,0,1))
    ax.set_yticks((-1,0,1))
    ax.set_zticks((-1,0,1))
    ax.set_xlabel('x')
    ax.set_ylabel('y')
    ax.set_zlabel('z')
    return
```

```
s = 2 # skip a few data points for display.
lw = 0.5 # linewidth of the arrows.
figsize = (3,3)

# Straight wire example
p, curr = currents_along_line(1,2,0.1)
ax = plt.figure(figsize=figsize).add_subplot(projection='3d')
ax.set_box_aspect(aspect=(1,1,1))
ax.quiver(p[0,::s],p[1,::s],p[2,::s],
          curr[0,::s],curr[1,::s],curr[2,::s],
          color='black',linewidth=lw)
tidy_ax(ax)
plt.title('Currents along line')
plt.tight_layout()
plt.savefig('fig_ch5_curr_line.pdf',bbox_inches='tight')
plt.show()
```

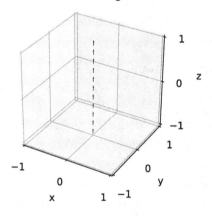

Figure 5.2

In the next two code blocks, we specify and visualize a steady current through a square loop with a side of length L and a ring of radius R. Just as in the previous code block, the current source points (x_s, y_s, z_s) on them are specified with an `np.arange()` function with a constant interval of dL (for a square) or d_phi (for a ring). The `np.zeros()` function is used to prepare the array of a known size.

For simplicity, the square and the ring are placed on the xy-plane with the current flow in a counter-clockwise direction when viewed from the positive z-axis. For a square loop, the length segment vector $\vec{dl}$ depends on which of the four sides is being considered. For example, along the top side of the loop, $\vec{dl}$ would be $-dx\hat{x}$, while along the right side, $\vec{dl}$ is $+dy\hat{y}$. We construct the square loop by specifying the position and the current vectors along the four separate sides, as implemented in the function currents_along_square(). Alternatively, we can utilize a rotation matrix. Once one side of the square is defined, we can rotate it by 90° about the origin (mathematically equivalent to switching x to $-y$ and y to x) and build another side of the square. The function currents_along_square_with_rotation_matrix() defines this rotation matrix R90 and applies it three times with a for-loop.

```python
# Code Block 5.2

def currents_along_square (I,L,dL):
    # Current distribution for a square loop
    # with side L at origin. Each side is defined separately.
    d = L/2
    vec = np.arange(-d,d,dL) # one side
    N = len(vec)
    p = np.zeros((3,4*N))
    curr = np.zeros((3,4*N))

    i = 0 # right side
    p[0,i*N:(i+1)*N] = d
    p[1,i*N:(i+1)*N] = vec
    curr[1,i*N:(i+1)*N] = I*dL

    i = 1 # top side
    p[0,i*N:(i+1)*N] = -vec
    p[1,i*N:(i+1)*N] = d
    curr[0,i*N:(i+1)*N] = -I*dL

    i = 2 # left side
    p[0,i*N:(i+1)*N] = -d
    p[1,i*N:(i+1)*N] = -vec
    curr[1,i*N:(i+1)*N] = -I*dL

    i = 3 # bottom side
    p[0,i*N:(i+1)*N] = vec
    p[1,i*N:(i+1)*N] = -d
    curr[0,i*N:(i+1)*N] = I*dL

    return p, curr

def currents_along_square_with_rotation_matrix (I,L,dL):
    # Square loop can be constructed by rotating
    # a single wire by 90 degrees repeatedly.

    # Rotation matrix (90 degrees around z-axis).
    R90 = np.array([[0,-1,0],[1,0,0],[0,0,1]])

    vec = np.arange(-L/2,L/2,dL) # one side
    N = len(vec)
```

```
        p = np.zeros((3,4*N))
        curr = np.zeros((3,4*N))

        # Specify the right side of the square.
        p[0] = L/2
        p[1,:N] = vec
        curr[1,:N] = 1

        # The other three sides are rotated versions of this.
        for i in range(3):
            p[:,(i+1)*N:(i+2)*N] = np.matmul(R90,p[:,i*N:(i+1)*N])
            curr[:,(i+1)*N:(i+2)*N] = np.matmul(R90,curr[:,i*N:(i+1)*N])
        curr = curr*I*dL
        return p, curr

# Square example
p, curr = currents_along_square(1,1.6,0.1)
ax = plt.figure(figsize=figsize).add_subplot(projection='3d')
ax.quiver(p[0,::s],p[1,::s],p[2,::s],
          curr[0,::s],curr[1,::s],curr[2,::s],
          color='black',linewidth=lw)
tidy_ax(ax)
ax.set_title('Currents along square')
plt.tight_layout()
plt.savefig('fig_ch5_curr_square.pdf',bbox_inches='tight')
plt.show()
```

In the case of a ring, we use a single variable ϕ (phi in the code block) to specify each source point as $(x_s, y_s) = (R\cos\phi, R\sin\phi)$, where the angle ϕ is measured from the positive x-axis in radians. The current flows in the counterclockwise direction as seen from the $+z$-axis, so $Id\vec{l} = -IRd\phi\sin\phi\hat{x} + IRd\phi\cos\phi\hat{y}$.

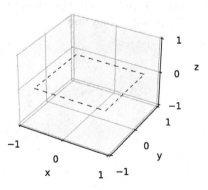

Currents along square

Figure 5.3

```
# Code Block 5.3

def currents_along_circle (I,R,d_phi):
    # Current distribution for a ring centered at the origin.
    # phi is the angle from the positive x-axis.
    # p = (x,y,0)
    # curr = I*d_phi * (-y,x,0)
    phi = np.arange(0,2*np.pi+d_phi,d_phi)
    N = len(phi)
    p = np.zeros((3,N))
    x, y = R*np.cos(phi), R*np.sin(phi)
    p[0], p[1] = x, y

    curr = np.zeros((3,N))
    curr[0], curr[1] = -y/R, x/R
    curr = curr*I*R*d_phi
    return p, curr

# Ring example
p, curr = currents_along_circle(1,1,np.pi/30)
ax = plt.figure(figsize=figsize).add_subplot(projection='3d')
ax.quiver(p[0,::s],p[1,::s],p[2,::s],
          curr[0,::s],curr[1,::s],curr[2,::s],
          color='black',linewidth=lw)
tidy_ax(ax)
ax.set_title('Currents along ring')
plt.tight_layout()
plt.savefig('fig_ch5_curr_ring.pdf',bbox_inches='tight')
plt.show()
```

Now that we have prepared three different cases (straight line, square loop, and ring) of current distributions in the above code blocks, let us work on preparing the observation points (x, y, z) at which the magnetic field will be calculated according to the Biot-Savart law. As in the previous chapters of working with the electric field, we will use

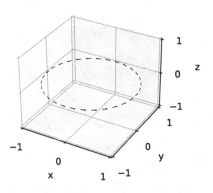

Figure 5.4

`np.meshgrid()` and `np.vstack()` functions to form an array of points in three dimensions. The code block below prepares and displays points in the cartesian coordinates.

```
# Code Block 5.4

# Prepare a set of points in 3D
# where the magnetic field will be evaluated.

def points_cartesian (d,delta):
    # (x,y,z) are between -d and d, with spacing of delta.
    v = np.arange(-d,d+delta,delta)
    x, y, z = np.meshgrid(v,v,v,indexing='ij')
    p = np.vstack((x.flatten(),y.flatten(),z.flatten()))
    return p

def plot_coordinates(p,title=''):
    fig = plt.figure(figsize=(3,3))
    ax = fig.add_subplot(projection='3d')
    ax.scatter(p[0],p[1],p[2],color='black')
    ax.set_xlabel('x')
    ax.set_ylabel('y')
    ax.set_zlabel('z')
    ax.set_xticks((-1,0,1))
    ax.set_yticks((-1,0,1))
    ax.set_zticks((-1,0,1))
    ax.set_title(title)
    plt.tight_layout()

p = points_cartesian(1,0.5)
plot_coordinates(p,title='Points in Cartesian Coordinate')
plt.savefig('fig_ch5_cartesian_coord.pdf',bbox_inches='tight')
plt.show()
```

The array p in the following code block stores the points (x, y, z) at which we will compute the magnetic fields. The magnetic field is generated from the current elements $I d\vec{l}$, which are located at (x_s, y_s, z_s). The current vectors and their positions are stored in the `curr` and `p_curr` arrays, respectively.

Points in Cartesian Coordinate

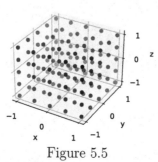

Figure 5.5

Because the two arrays, p and p_curr, specify two different sets of positions, we cannot calculate displacement vectors by simply subtracting them, as p - p_curr will yield an error. Hence, we create a size-matched temporary array p_tmp to find the difference between (x, y, z) and (x_s, y_s, z_s), and this displacement vector is assigned to an array r.

The cross product between the displacement and the current vectors can be calculated with a built-in NumPy np.cross() function, or we can create our own version of cross_product(). The array dB stores the magnetic fields due to individual current elements. Hence, by taking the sum of dB along an appropriate axis, we can calculate the total magnetic field at each observation location specified in p.

What happens if some of the source and observation points coincide? We would encounter an error since the denominator dist3 would be zero. Thus, we will try to avoid such an error, by carefully choosing the locations of the source and observation points and ensuring that they do not overlap. Although not implemented here, there could be a screening to find and disregard points for which dist3 is zero (or very small).

```python
# Code Block 5.5

def cross_product (A, B):
    # A and B: 3 by N arrays
    # This function is the same as np.cross(A.T,B.T).T
    V = np.zeros(A.shape)
    V[0] = A[1]*B[2] - A[2]*B[1]
    V[1] = A[2]*B[0] - A[0]*B[2]
    V[2] = A[0]*B[1] - A[1]*B[0]
    return V

def get_magnetic_field (p,p_curr,curr):

    # p: observation points
    # p_curr: position coordinate of current
    # curr: vector of current (times dl)

    assert p_curr.shape[1]==curr.shape[1]

    mu0 = 4*np.pi*(10**-7)
    M = p.shape[1]
    N = p_curr.shape[1]

    B = np.zeros(p.shape)

    for i in range(M):
```

```
    p_tmp = np.reshape(p[:,i],(3,1)).dot(np.ones((1,N)))
    r = p_tmp - p_curr # displacement vector: 3 by N
    dist3 = np.sqrt(np.sum(r**2,axis=0))**3
    # Warning: dist3 could be very small or zero.
    dB = cross_product(curr,r) / dist3

    # Cross-product can be also calculated with a built-in
    # function in the numpy module, np.cross.
    #dB = np.cross(curr.T,r.T).T / dist3

    B[:,i] = np.nansum(dB,axis=1)
B = (muO/(4*np.pi))*B

# Magnitude of B can be calculated by:
#B_mag = np.sqrt(np.sum(B**2,axis=1))

return B
```

Let us put all of the above together. In the next code blocks, we examine three different cases of magnetic fields. The first example is the magnetic field generated by a straight wire. The magnetic field may be described as "flowing" or "swirling" around the wire according to the right-hand rule. If you point the thumb of your right hand along the current, the naturally curving direction of the other four fingers shows the direction of the magnetic field. The magnetic field gets weaker farther away from the wire.

```
# Code Block 5.6

# Plot the magnetic field in three dimensions.

def plot_magnetic_field_3d (p,p_curr,curr,B,title=''):

    fig = plt.figure(figsize=(3,3))
    ax = fig.add_subplot(projection='3d')

    ax.plot(p_curr[0],p_curr[1],p_curr[2],
            color='gray',linewidth=10,alpha=0.5)
    ax.quiver(p[0],p[1],p[2],B[0],B[1],B[2],
              color='black',linewidth=0.5,
              length=0.2,alpha=0.5,normalize=True)
    ax.set_title(title)
    ax.set_xlabel('x')
    ax.set_ylabel('y')
    ax.set_zlabel('z')
```

```
    ax.set_xticks((-1,0,1))
    ax.set_yticks((-1,0,1))
    ax.set_zticks((-1,0,1))
    plt.tight_layout()
```

```
# Code Block 5.7

# Visualize magnetic field around:
# straight wire, square loop, ring.
# The arrow lengths are not proportional to the magnitude.

# Straight wire
I, L, dL = 1, 5, 0.1
p_curr, curr = currents_along_line(I,L,dL)
p = points_cartesian (0.8,0.4)
B = get_magnetic_field(p,p_curr,curr)
plot_magnetic_field_3d(p,p_curr,curr,B,title='Line')
plt.savefig('fig_ch5_B_line.pdf')
plt.show()

# Square
I, L, dL = 1, 1, 0.01
p_curr, curr = currents_along_square(I,L,dL)
p = points_cartesian (0.7,0.2)
B = get_magnetic_field(p,p_curr,curr)
plot_magnetic_field_3d(p,p_curr,curr,B,title='Square')
plt.savefig('fig_ch5_B_square.pdf')
plt.show()

# Ring
I, R = 1, 0.5
p_curr, curr = currents_along_circle(I,R,np.pi/100)
p = points_cartesian (0.7,0.2)
B = get_magnetic_field(p,p_curr,curr)
plot_magnetic_field_3d(p,p_curr,curr,B,title='Ring')
plt.savefig('fig_ch5_B_ring.pdf')
plt.show()
```

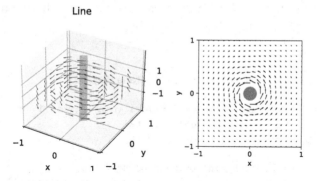

Figure 5.6

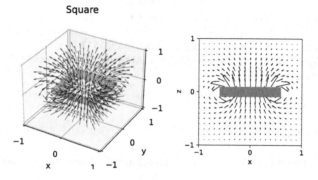

Figure 5.7

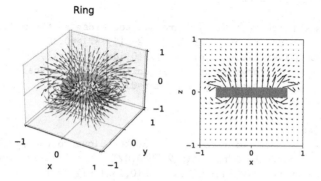

Figure 5.8

The plots of the two-dimensional cross-sections of the magnetic fields were generated by the following code blocks.

```python
# Code Block 5.8

# Define observation points on a 2D-plane (xy or xz).

def points_cartesian_xy (xmin=-1,xmax=1,ymin=-1,ymax=1,delta=0.1):
    x = np.arange(xmin,xmax+delta,delta)
    y = np.arange(ymin,ymax+delta,delta)
    z = 0
    xv,yv,zv = np.meshgrid(x,y,z,indexing='ij')
    p = np.vstack((xv.flatten(),yv.flatten(),zv.flatten()))
    return p

def points_cartesian_xz (xmin=-1,xmax=1,zmin=-1,zmax=1,delta=0.1):
    x = np.arange(xmin,xmax+delta,delta)
    z = np.arange(zmin,zmax+delta,delta)
    y = 0
    xv,yv,zv = np.meshgrid(x,y,z,indexing='ij')
    p = np.vstack((xv.flatten(),yv.flatten(),zv.flatten()))
    return p
```

```python
# Code Block 5.9

# Plot the magnetic field in 2D.

def plot_magnetic_field_2d (p,p_curr,curr,B,view='xy',title=''):

    fig = plt.figure(figsize=(3,3))
    scale = 7*(10**(-6))
    lw = 0.5
    ax = fig.add_subplot()
    if view=='xy':
        # Consider points that are not too close to the z-axis.
        idx = np.where(np.sqrt(p[0]**2+p[1]**2)>=0.15)
        ax.quiver(p[0,idx],p[1,idx],B[0,idx],B[1,idx],
                  angles='xy',scale_units='xy',scale=scale,
                  linewidth=lw,zorder=0)
        ax.set_ylabel('y')

    if view=='xz':
        # Consider points that are not too close to the xy-plane.
        idx = np.where(np.abs(p[2])>=0.02)
        ax.quiver(p[0,idx],p[2,idx],B[0,idx],B[2,idx],
                  angles='xy',scale_units='xy',scale=scale,
                  linewidth=lw,zorder=0)
        ax.set_ylabel('z')
```

```
    ax.set_title(title)
    ax.axis('square')
    ax.set_xlabel('x')
    ax.set_xticks((-1,0,1))
    ax.set_yticks((-1,0,1))
    ax.set_xlim((-1,1))
    ax.set_ylim((-1,1))


lw = 15 # linewidth of the wire

# wire
I, L, dL = 1, 5, 0.1
p_curr, curr = currents_along_line(I,L,dL)
p = points_cartesian_xy()
B = get_magnetic_field(p,p_curr,curr)
plot_magnetic_field_2d(p,p_curr,curr,B,view='xy',title='')
plt.scatter([0],[0],s=400,color='gray') # wire
plt.tight_layout()
plt.savefig('fig_ch5_B_line_2d.pdf',bbox_inches='tight')
plt.show()

# square loop
I, L, dL = 1, 1, 0.01
p_curr, curr = currents_along_square(I,L,dL)
p = points_cartesian_xz (zmin=-0.9,zmax=0.9,delta=0.1)
B = get_magnetic_field(p,p_curr,curr)
plot_magnetic_field_2d(p,p_curr,curr,B,view='xz',title='')
plt.plot([-L/2,L/2],[0,0],linewidth=lw,color='gray',zorder=1)
plt.tight_layout()
plt.savefig('fig_ch5_B_square_2d.pdf',bbox_inches='tight')
plt.show()

# ring
I, R = 1, 0.6
p_curr, curr = currents_along_circle(I,R,np.pi/100)
p = points_cartesian_xz (zmin=-0.9,zmax=0.9,delta=0.1)
B = get_magnetic_field(p,p_curr,curr)
plot_magnetic_field_2d(p,p_curr,curr,B,view='xz',title='')
plt.plot([-R,R],[0,0],linewidth=lw,color='gray',zorder=1)
plt.tight_layout()
plt.savefig('fig_ch5_B_ring_2d.pdf',bbox_inches='tight')
plt.show()
```

As shown above, the Biot-Savart law allows us to calculate magnetic fields generated by a steady electric current in an arbitrary shape, by dividing it into small current segments and adding up their individual contributions. We have illustrated the Biot-Savart law with three examples. The implication of the cross-product embedded inside the Biot-Savart law is that the magnetic field "circulates" around currents, and its circulating direction can be described with a right-hand rule.

For a current through a loop, whether it is a square or a ring, the right-hand rule reveals the direction of the magnetic field inside and outside the loop. If the current flows counterclockwise when the loop is viewed from the top, the magnetic field inside the loop points toward the top. When multiple loops are stacked together in the form of a long coil, also known as a solenoid, the magnetic field from each loop reinforces one another, producing a strong and uniform magnetic field inside.

For a few simple cases, the integral of the Biot-Savart law can be calculated analytically, producing an exact formula. We can check the validity of our numerical calculations with these formulae. In the case of a straight wire of length L with current I flowing in $+z$-direction, the magnetic field at a distance r away from its midpoint is given by

$$\vec{B}_{\text{wire}} = \frac{\mu_0 I L}{2\pi r \sqrt{4r^2 + L^2}}(-\sin\phi\hat{x} + \cos\phi\hat{y}).$$

For a square loop with side L and a circular ring with radius R, lying on the xy-plane and having the current flow I in a counter-clockwise direction as viewed from the $+z$-axis, the magnetic fields at distance r above their centers are

$$\vec{B}_{\text{square}} = \frac{\mu_0 I L^2}{2\pi\left(\left(\frac{L}{2}\right)^2 + r^2\right)\sqrt{\frac{L^2}{2} + r^2}}\hat{z}$$

and

$$\vec{B}_{\text{ring}} = \frac{\mu_0 I R^2}{2\left(R^2 + r^2\right)^{3/2}}\hat{z}.$$

We leave the verification of these results to our readers as an exercise.

Before we calculate the magnetic fields in two different ways (analytically and numerically) and compare them, let us discuss how we might compare two vectors in general. As shown in the following figure, the difference between two vectors, $\vec{A}$ and $\vec{B}$, is also a vector. Therefore, we may examine the magnitude of the difference vector with reference to one of the original vectors and calculate the fraction of difference, $\frac{|\vec{B}-\vec{A}|}{|\vec{A}|}$. Furthermore, we can calculate the normalized dot product between two vectors $\frac{\vec{A}\cdot\vec{B}}{|\vec{A}||\vec{B}|}$ and examine how aligned they are. If these two vectors are closely matched, the normalized dot product will be close to 1.

```python
# Code Block 5.10

# Quantify the difference between two vectors.

fig = plt.figure(figsize=(3,3))
A = np.array([0.5, 0.8])
B = np.array([0.55, 0.75])
C = B-A
plt.quiver(0,0,A[0],A[1],
           angles='xy',scale_units='xy',scale=1,color='#000000')
plt.quiver(0,0,B[0],B[1],
           angles='xy',scale_units='xy',scale=1,color='#CCCCCC')
plt.quiver(A[0],A[1],C[0],C[1],
           angles='xy',scale_units='xy',scale=1,color='#808080')
plt.legend(('A','B','B-A'))
plt.xticks((0,0.5,1))
plt.yticks((0,0.5,1))
plt.xlim((-0.1,1.1))
plt.ylim((-0.1,1.1))
plt.axis('square')
plt.savefig('fig_ch5_vec_diff.pdf')
plt.show()

def compare_two_vectors (V_exact, V_apprx):
    norm_exact = np.sqrt(np.sum(V_exact**2))
    norm_apprx = np.sqrt(np.sum(V_apprx**2))
    diff = V_exact - V_apprx
    norm_diff = np.sqrt(np.sum(diff**2))
    print("Difference in magnitudes: %4.3f percent"
          %(norm_diff/norm_exact*100))

    dot_prod = np.sum(V_exact*V_apprx)/(norm_exact*norm_apprx)
    print("Difference in direction (Norm. Dot Product): %4.3f"
          %dot_prod)
    print('')

print('Example case: comparing two vectors')
compare_two_vectors(A,B)
```

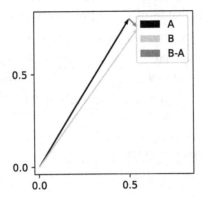

Figure 5.9

```
Example case: comparing two vectors
Difference in magnitudes: 7.495 percent
Difference in direction (Norm. Dot Product): 0.997
```

The following code block uses this approach of comparing two vectors. We calculate and compare the magnetic fields calculated from an exact analytical formula and a numerical method. As expected, the two vectors are well matched, and the agreement improves as we consider even smaller current segments (dL in the code). The analytical formulae give us the exact values, but they are applicable only for special cases with simple shapes and specific observation points (e.g., the magnetic field at the center of a circular ring). In contrast, the numerical method can be used for any arbitrarily shaped wire at an arbitrarily placed observation point.

```python
# Code Block 5.11

# Comparison of analytical and numerical solutions.

mu0 = 4*np.pi*(10**-7)

print("Comparison 1: Straight wire")
print("B at (0.2,0,0) due to a line of L = 5 with I = 2")
L, I, r = 5, 2, 0.2
phi = 0 # Because the observation point is on the x-axis.

# Analytical solution.
Bx = (-np.sin(phi))/(r*np.sqrt((4*r**2)+L**2))*L
By = (+np.cos(phi))/(r*np.sqrt((4*r**2)+L**2))*L
```

```
Bz = 0
B_wire_exact = np.array([[Bx],[By],[Bz]])*(mu0*I)/(2*np.pi)
B_wire_exact = B_wire_exact

# Numerical solution.
p = np.array([[r],[0],[0]])
p_curr, curr = currents_along_line(I,L,L*0.05)
B_wire_approx = get_magnetic_field(p,p_curr,curr)

compare_two_vectors (B_wire_exact,B_wire_approx)

print("Comparison 2: Square loop")
print("B at (0,0,0.3) due to a square of L = 1 with I = 2")
L, I, r = 1, 2, 0.3

# Analytical solution
Bx, By = 0, 0
Bz = (L**2)/((((L/2)**2)+(r**2))*np.sqrt(((L**2)/2+(r**2))))
B_square_exact = np.array([[Bx],[By],[Bz]])*(mu0*I)/(2*np.pi)

# Numerical solution
p = np.array([[0],[0],[r]])
p_curr, curr = currents_along_square(I,L,L*0.05)
B_square_approx = get_magnetic_field(p,p_curr,curr)

compare_two_vectors (B_square_exact,B_square_approx)

print("Comparison 3: Ring")
print("B at (0,0,0.3) due to a ring of R = 0.5 with I = 2")
R, I, r = 0.5, 2, 0.3

# Analytical calculation for the ring.
Bx, By = 0, 0
Bz = (mu0*I*R**2)/(2*(R**2+r**2)**(3/2))
B_ring_exact = np.array([[Bx],[By],[Bz]])

# Numerical calculation for the ring.
p = np.array([[0],[0],[r]])
p_curr, curr = currents_along_circle(I,R,np.pi/10)
B_ring_approx = get_magnetic_field(p,p_curr,curr)

compare_two_vectors (B_ring_exact,B_ring_approx)
```

Comparison 1: Straight wire
B at (0.2,0,0) due to a line of L = 5 with I = 2
Difference in magnitudes: 4.026 percent
Difference in direction (Norm. Dot Product): 1.000

```
Comparison 2: Square loop
B at (0,0,0.3) due to a square of L = 1 with I = 2
Difference in magnitudes: 0.061 percent
Difference in direction (Norm. Dot Product): 1.000

Comparison 3: Ring
B at (0,0,0.3) due to a ring of R = 0.5 with I = 2
Difference in magnitudes: 5.831 percent
Difference in direction (Norm. Dot Product): 1.000
```

5.2 AMPERE'S LAW

In the previous chapter, we discussed Gauss's law, which states that
the surface integral of the electric field is related to the total amount
of charge enclosed within the boundary of the integral (also known as
a Gaussian surface). Similarly, the line integral of the magnetic field
over a closed path or contour (also known as an Amperian loop) is related to the total amount of enclosed current, as we will discuss in this
chapter. As a starting example, consider two current-carrying wires and
various closed paths of different shapes (circle or square) at different locations. In the following code block, we reuse the previously defined function, currents_along_line(), to specify two wires along the z-axis. We
also reuse currents_along_square() and currents_along_circle(),
to specify the closed paths at different positions on the xy-plane.

```
# Code Block 5.12

# Calculate the magnetic field from two current wires.

L, dL = 100, 0.1

# You may experiment with these values.
I1, x01, y01 = +1, -0.65, -0.15 # wire 1
I2, x02, y02 = -1, +0.65, +0.15 # wire 2

p_curr1, curr1 = currents_along_line(I1,L,dL,x0=x01,y0=y01)
p_curr2, curr2 = currents_along_line(I2,L,dL,x0=x02,y0=y02)

# Calculate the total magnetic fields due to two currents.
p = points_cartesian_xy (xmin=-2,xmax=2,ymin=-2,ymax=2,delta=0.25)
B1 = get_magnetic_field(p, p_curr1, curr1)
B2 = get_magnetic_field(p, p_curr2, curr2)
B = B1 + B2

# Define 5 different closed paths for line integral.
```

```
C1, _ = currents_along_square (I=0, L=1.2, dL=0.1)
C1[0] = C1[0]-0.7

C2, _ = currents_along_circle (I=0, R=0.5, d_phi=np.pi/100)
C2[0] = C2[0]-0.4

C3, _ = currents_along_square (I=0, L=0.8, dL=0.1)
C3[0], C3[1] = C3[0]+0.9, C3[1]+1.2,

C4, _ = currents_along_circle (I=0, R=0.5, d_phi=np.pi/100)
C4[0] = C4[0]+1.5

C5, _ = currents_along_circle (I=0, R=2, d_phi=np.pi/100)

# Plot.
plt.figure(figsize=(6,6))

def get_marker (I):
    # Marker convention.
    # "." for a direction pointing out of the page.
    # "x" for a direction pointing into the page.
    if I > 0:
        marker = '.'
    else:
        marker = 'x'
    return marker
m1 = get_marker(I1)
m2 = get_marker(I2)

plt.scatter(p_curr1[0,0],p_curr1[1,0],color='gray',s=200)
plt.scatter(p_curr2[0,0],p_curr2[1,0],color='gray',s=200)
plt.scatter(p_curr1[0,0],p_curr1[1,0],color='white',s=80,marker=m1)
plt.scatter(p_curr2[0,0],p_curr2[1,0],color='white',s=80,marker=m2)
plt.quiver(p[0],p[1],B[0],B[1],color='black',
          angles='xy',scale_units='xy',scale=3*10**(-6))

lw = 2
marg = 0.05
fs = 14 # fontsize
plt.plot(C1[0],C1[1],color='gray',linewidth=lw)
plt.plot((C1[0,0],C1[0,-1]),(C1[1,0],C1[1,-1]),
        color='gray',linewidth=lw) # close up the square.
plt.plot(C2[0],C2[1],color='gray',linewidth=lw)
plt.plot(C3[0],C3[1],color='gray',linewidth=lw)
plt.plot((C3[0,0],C3[0,-1]),(C3[1,0],C3[1,-1]),
        color='gray',linewidth=lw) # close up the square.
plt.plot(C4[0],C4[1],color='gray',linewidth=lw)
plt.plot(C5[0],C5[1],color='gray',linewidth=lw)
plt.text(C1[0,24]+marg,C1[1,24]+marg,r"$C_1$",fontsize=fs)
```

```
plt.text(C2[0,100]+marg,C2[1,100]+marg,r"$C_2$",fontsize=fs)
plt.text(C3[0,0]+marg,C3[1,0]+marg,r"$C_3$",fontsize=fs)
plt.text(C4[0,50]+marg,C4[1,50]+marg,r"$C_4$",fontsize=fs)
plt.text(C5[0,100]+marg,C5[1,100]+marg,r"$C_5$",fontsize=fs)
plt.xlabel('x')
plt.ylabel('y')
plt.xlim((-2,2))
plt.ylim((-2,2))
plt.xticks((-2,0,2))
plt.yticks((-2,0,2))
plt.title('Currents, magnetic fields, and closed paths')
plt.axis('square')
plt.savefig('fig_ch5_curr_B_paths.pdf',bbox_inches='tight')
plt.show()
```

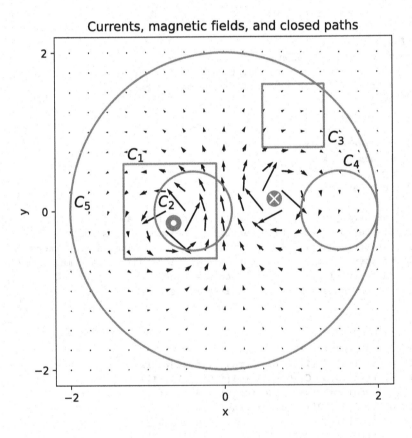

Figure 5.10

Once a vector field is defined, a line integral over a path is calculated by integrating the dot product between the infinitesimal path segment and the field vector at each point. The mathematical notation is given by $\oint_{\text{Contour}} \vec{B} \cdot \vec{dl}$ for the line integral of the magnetic field $\vec{B}$ over a closed contour.

The following code block contains a function for performing the line integral over five different contours examples. As a convention, we will go around the closed path in a counter-clockwise direction. The for-loop at the bottom of the code prints out the calculated values that are normalized with respect to μ_0. When the contour includes only one wire, the line integral has a value of 1. When the contour includes either none or both wires carrying equal-and-opposite currents, the line integral is zero.

```python
# Code Block 5.13

# Perform line integrals over different closed paths.

def line_integral_vector_field (vfield, p):
    dx, dy = np.diff(p[0]), np.diff(p[1])
    vx, vy = vfield[0][:-1], vfield[1][:-1]
    val0 = np.sum(vx*dx+vy*dy)
    vx, vy = vfield[0][1:], vfield[1][1:]
    val1 = np.sum(vx*dx+vy*dy)
    line_integral = (val0 + val1)/2
    return line_integral

B1_C1 = get_magnetic_field(C1, p_curr1, curr1)
B1_C2 = get_magnetic_field(C2, p_curr1, curr1)
B1_C3 = get_magnetic_field(C3, p_curr1, curr1)
B1_C4 = get_magnetic_field(C4, p_curr1, curr1)
B1_C5 = get_magnetic_field(C5, p_curr1, curr1)

B2_C1 = get_magnetic_field(C1, p_curr2, curr2)
B2_C2 = get_magnetic_field(C2, p_curr2, curr2)
B2_C3 = get_magnetic_field(C3, p_curr2, curr2)
B2_C4 = get_magnetic_field(C4, p_curr2, curr2)
B2_C5 = get_magnetic_field(C5, p_curr2, curr2)

lint_C1 = line_integral_vector_field(B1_C1+B2_C1,C1)
lint_C2 = line_integral_vector_field(B1_C2+B2_C2,C2)
lint_C3 = line_integral_vector_field(B1_C3+B2_C3,C3)
lint_C4 = line_integral_vector_field(B1_C4+B2_C4,C4)
lint_C5 = line_integral_vector_field(B1_C5+B2_C5,C5)

lint = np.array([lint_C1,lint_C2,lint_C3,lint_C4,lint_C5])

for i,v in enumerate(lint/mu0):
```

```
    print("Line integral over C%d = %+5.4f"%(i+1,v))
```

```
Line integral over C1 = +0.9921
Line integral over C2 = +0.9998
Line integral over C3 = -0.0104
Line integral over C4 = +0.0000
Line integral over C5 = -0.0000
```

As a further experiment, let's calculate the line integral while changing the amount of current through one of the wires in the following code block. The top figure shows the line integral over four closed contours of C_1, C_2, C_3, and C_4. When the line integrals of the magnetic field over the paths C_1 and C_2 are compared, they return the same value of $\oint_{C_1} \vec{B} \cdot d\vec{l} = \oint_{C_2} \vec{B} \cdot d\vec{l} = \mu_0$, within numerical precision. This is also true for C_3 and C_4, where $\oint_{C_3} \vec{B} \cdot d\vec{l} = \oint_{C_4} \vec{B} \cdot d\vec{l} = 0$.

What makes the line integral over C_1 and C_2 identical is the fact that both contours contain the same, fixed amount of current through a single wire. The contours C_3 and C_4 do not contain any current, so the line integrals over them are zero. This observation indicates that a contour's position and shape do not matter for the line integral. What matters is the amount of current enclosed within the path. Furthermore, the current flowing outside the closed path does not affect the value of the line integral. In the case of C_5, which encloses both wires, the line integral depends on the total amount of the enclosed current, as illustrated by the bottom figure. Since I2 was varied linearly, the calculated line integral values also show linear increases with a slope equal to μ_0.

```
# Code Block 5.14

# Explore how the line integral varies as I2 is varied (I1 fixed).

I1 = 1
I2_range = np.arange(-2.5,0.6,0.25)
integral = np.zeros((len(I2_range),5))

# Calculate B1 due to I1
p_curr1, curr1 = currents_along_line(I1,L,dL,x0=x01,y0=y01)
B1_C1 = get_magnetic_field(C1, p_curr1, curr1)
B1_C2 = get_magnetic_field(C2, p_curr1, curr1)
B1_C3 = get_magnetic_field(C3, p_curr1, curr1)
B1_C4 = get_magnetic_field(C4, p_curr1, curr1)
```

```
B1_C5 = get_magnetic_field(C5, p_curr1, curr1)

for i, I2 in enumerate(I2_range):

    # Calculate B2 due to I2
    p_curr2, curr2 = currents_along_line(I2,L,dL,x0=x02,y0=y02)
    B2_C1 = get_magnetic_field(C1, p_curr2, curr2)
    B2_C2 = get_magnetic_field(C2, p_curr2, curr2)
    B2_C3 = get_magnetic_field(C3, p_curr2, curr2)
    B2_C4 = get_magnetic_field(C4, p_curr2, curr2)
    B2_C5 = get_magnetic_field(C5, p_curr2, curr2)

    # Calculate line integral of B1+B2 over different paths.
    lint_C1 = line_integral_vector_field(B1_C1+B2_C1,C1)
    lint_C2 = line_integral_vector_field(B1_C2+B2_C2,C2)
    lint_C3 = line_integral_vector_field(B1_C3+B2_C3,C3)
    lint_C4 = line_integral_vector_field(B1_C4+B2_C4,C4)
    lint_C5 = line_integral_vector_field(B1_C5+B2_C5,C5)

    lint = np.array([lint_C1,lint_C2,lint_C3,lint_C4,lint_C5])
    integral[i,:] = lint / mu0

fig = plt.figure(figsize=(4,3))
plt.scatter(I2_range,integral[:,0],color='#CCCCCC',label='$C_1$',s=90)
plt.scatter(I2_range,integral[:,1],color='#000000',label='$C_2$',s=10)
plt.scatter(I2_range,integral[:,2],color='#808080',label='$C_3$',s=90)
plt.scatter(I2_range,integral[:,3],color='#CCCCCC',label='$C_4$',s=10)
plt.title(r'$\oint Bdl$ over $C_1$, $C_2$, $C_3$, $C_4$')
plt.xlabel('$I_2$')
plt.xticks((-2,-1,0))
plt.yticks((0,0.5,1))
plt.ylabel(r'$\frac{1}{\mu_0}\oint Bdl$')
plt.legend(framealpha=1)
plt.savefig('fig_ch5_diff_paths_1.pdf',bbox_inches='tight')
plt.show()

fig = plt.figure(figsize=(4,3))
plt.scatter(I1+I2_range,integral[:,4],label='$C_5$',color='black')
plt.plot((-1.5,1.5),(-1.5,1.5),color='#AAAAAA',
         label='Line with a slope of 1')
plt.title('$\oint Bdl$ over $C_5$')
plt.xlabel('$I_1+I_2$')
plt.yticks((-1,0,1))
plt.xticks((-1,0,1))
plt.ylabel(r'$\frac{1}{\mu_0}\oint Bdl$')
plt.legend(framealpha=1)
plt.savefig('fig_ch5_diff_paths_2.pdf',bbox_inches='tight')
plt.show()
```

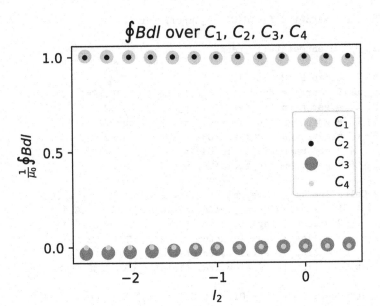

Figure 5.11

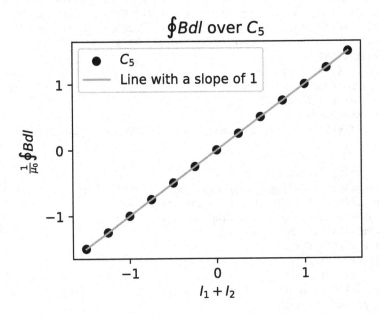

Figure 5.12

Our observation so far has revealed the following:

$$\oint_{\text{Contour}} \vec{\mathbf{B}} \cdot \vec{dl} = \mu_0 \sum I,$$

where $\sum I$ is the net current inside the contour or the path. This result is known as Ampere's law. We note that the direction of the current matters. If two wires carry an equal magnitude of currents but in opposite directions, $\sum I$ will be zero, not twice the current each wire carries, as illustrated by the path C_5.

5.3 CAUTIONARY EXAMPLE

Let us discuss a short cautionary example illustrating how numerical imprecision may creep into our calculations. In the following code block, we prepare three cases where multiple currents flow in parallel. Currents are spaced dx apart and arranged symmetrically along the x-axis. Because of the symmetry, the magnetic field at the origin is expected to be zero, but even in these simple cases with 4 or 6 current sources, we obtain non-zero, although small, magnetic fields along the y-axis in B_total.

We also note that when np.arange() is called to generate an array of equally spaced numbers between -1 and 1, the midpoint is not an exact zero but carries a small numerical value. Such loss of numerical precision is not unique to Python. Other computational platforms or programming languages have the same issue since the floating points are represented with a fixed number of bits.

When calculating electric or magnetic fields, we sometimes deal with a case where the source of the field (i.e., charges or currents, respectively) may be placed rather close to the point at which the field is being calculated. Because Coulomb's or Biot-Savart laws involve the division by some power of distances, a small numerical imprecision in the denominator may be amplified and generate imprecise results. We will see such an example in the next section.

```
# Code Block 5.15

# Calculate magnetic field from multiple currents.

def calculate_B_multiple_currents(x0_range):
    I = 1
    L, dL = 10, 0.1
    p = np.array([[0],[0],[0]])
```

```
    B_tot = np.zeros((3,1))
    for x0 in x0_range:
        p_curr, curr = currents_along_line(I,L,dL,x0=x0,y0=0)
        B = get_magnetic_field(p, p_curr, curr)
        B_tot = B_tot + B
    return B_tot

def plot_multiple_currents(x0_range):
    plt.figure(figsize=(5,1))
    for x0 in x0_range:
        plt.scatter(x0,0,color='black',s=200)
        plt.scatter(x0,0,color='white',s=80,marker='.')
    plt.yticks((-1,0,1))
    plt.xticks((-1,0,1))
    plt.ylim((-0.5,0.5))
    plt.xlim((-1.1,1.1))
    plt.xticks(None)
    plt.tight_layout()

# In all cases, we are supposed to see B_total = [0,0,0],
# but we may run into numerical precision issues.

for i in (1,3,5):
    dx = 0.2
    x0_range = np.arange(-i*dx/2,i*dx/2+dx,dx)
    B_total = calculate_B_multiple_currents(x0_range)
    print("B in the middle =",B_total.T[0])
    plot_multiple_currents(x0_range)
    plt.savefig('fig_ch5_mult_current_%d.pdf'%(i+1),
                bbox_inches='tight')
    plt.show()

# Even the built-in function np.arange returns
# a small non-zero value at the position of zero.
print('Print: np.arange(-1,1.1,0.1)')
print(np.arange(-1,1.1,0.1))
```

B in the middle = [0. 0. 0.]

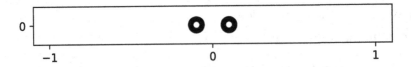

Figure 5.13

B in the middle = [0.00000000e+00 -1.05879118e-21 0.00000000e+00]

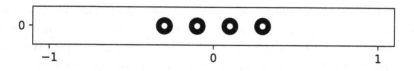

Figure 5.14

B in the middle = [0.00000000e+00 2.01170325e-21 0.00000000e+00]

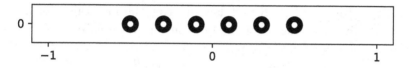

Figure 5.15

```
Print: np.arange(-1,1.1,0.1)
[-1.00000000e+00 -9.00000000e-01 -8.00000000e-01 -7.00000000e-01
 -6.00000000e-01 -5.00000000e-01 -4.00000000e-01 -3.00000000e-01
 -2.00000000e-01 -1.00000000e-01 -2.22044605e-16  1.00000000e-01
  2.00000000e-01  3.00000000e-01  4.00000000e-01  5.00000000e-01
  6.00000000e-01  7.00000000e-01  8.00000000e-01  9.00000000e-01
  1.00000000e+00]
```

5.4 EXAMPLE: THICK WIRE

Our discussions so far have dealt with electric current through a thin wire of negligible thickness. Here we will consider a wire with non-zero thickness and a current density $\vec{J}$ in it. The direction of $\vec{J}$ corresponds to the direction of the current flow, and its magnitude is equal to the amount of current per cross-sectional area. Hence, $\vec{J}$ is a vector quantity.

The total amount of the current I can be determined by adding up the flow of charges at each point in space.

$$I = \iint_{\text{Area}} \vec{J} \cdot \hat{n} da,$$

where $\hat{n}$ is a unit vector normal to the cross-sectional area da. In other words, current is the flux of current density over an area.

The following figure compares the currents through wires with zero and with non-zero thickness. For the zero thickness case, we apply the Biot-Savart law by integrating over the length of a wire. For the case with current in a wire with finite thickness, the current element $I\vec{dl}$ has to be considered as $(\vec{J} \cdot \hat{n}da)\vec{dl}$, and hence we have to integrate over the differential cross-sectional areas as well as the length segments of the wire.

$$\vec{B}(x,y,z) = \frac{\mu_0}{4\pi} \int_{\text{length}} \iint_{\text{cross-sectional area}} \frac{(\vec{J} \cdot \hat{n}da)\vec{dl} \times \vec{r}}{r^3}.$$

In other words, the Biot-Savart law involves the integrals over da and $\vec{dl}$, essentially becoming a volume integration over a wire.

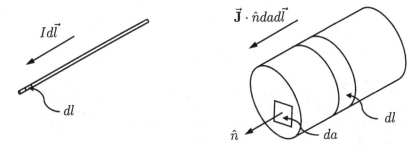

Figure 5.16

Consider a wire whose cross-section is a disc with a radius R and a total current I. To make the problem more interesting, imagine that this wire has a non-uniform current density, such that more currents flow closer to the edge of the wire. More specifically, let's assume that the current density is proportional to r^2 where r is the distance from the center of the circular cross-section of the wire. Hence, $\vec{J} = \alpha r^2 \hat{n}$, where α is a proportionality constant and $\hat{n}$ is in the same direction as $\vec{J}$. The r^2 dependence is just an arbitrary example. We could have chosen the current density to be proportional to r or r^3. Such non-uniform current density could be due to non-uniform physical properties of a wire, such as temperature gradient or different material compositions.

We can determine the expression for α by calculating the total current in the polar coordinates: $I = \iint_{\text{Area}} \vec{J} \cdot \hat{n}da = \int_0^R \int_0^{2\pi} \vec{J} \cdot \hat{n}rdrd\phi = \int_0^R \int_0^{2\pi} \alpha r^3 drd\phi = \frac{\alpha\pi R^4}{2}$. Hence, $\alpha = \frac{2I}{\pi R^4}$. The following code block verifies the expression for α with the **sympy** module.

```
# Code Block 5.16

# Find the proportionality constant, alpha,
# for current density J = alpha*r^2.

import sympy as sym
I, r, R, phi = sym.symbols('I r R phi')
alpha_symbolic = I/sym.integrate(r**3,(r,0,R),(phi,0,2*sym.pi))
display(alpha_symbolic)
```

$$\frac{2I}{\pi R^4}$$

Now we have a well-defined expression for current density:

$$\vec{J}(r) = \frac{2I}{\pi R^4} r^2 \hat{z},$$

where we assume that the current is flowing along the z-direction, so that the cross-section of the wire on the xy-plane is a disc with radius R.

The quantity Idl necessary to calculate the magnetic field with the Biot-Savart law is $Idl = \vec{J} \cdot \hat{z} dadl = \alpha r^2 dV$, where dV is a small volume element in the wire. In the following code block, the function r2_curr_from_p_curr() returns curr for given p_curr (a vector of positions in the wire), I, R, L, and the differential volume element dV. Because p_curr specifies the (x, y, z) coordinate values, it is straightforward to calculate the distance r from the center of the wire as r = np.sqrt(p_curr[0]**2 + p_curr[1]**2).

To systematically specify an array of sample positions in the wire (p_curr) and their corresponding volume elements (dV), we can take several different approaches. In the function curr_density_r2_polar(), the points are sampled in the polar coordinates along r and ϕ with $dV = rdrd\phi$. If the points are sampled along the cartesian coordinates as shown in curr_density_r2_cartesian(), $dV = dxdydz$ (or dV = (dR**2)*dL). Finally, in curr_density_r2_random(), we pick N random points inside the volume of the wire. Hence, the differential volume element at each point is approximately given by $dV = V/N = \pi R^2 L/N$.

```
# Code Block 5.17

# Consider a wire of length L with a finite thickness.
# It has a circular cross section with radius R.
```

```
# The current density is not uniform (r-dependence).

# Define current vector at each point in the array p_curr.
def r2_curr_from_p_curr (I,R,L,p_curr,dV):
    # Specify r^2 current from the p_curr array.
    r = np.sqrt(p_curr[0]**2 + p_curr[1]**2)
    N = len(r) # Number of sample points
    alpha = (2*I)/(np.pi*(R**4))
    curr = np.zeros((3,N))
    curr[2] = alpha*(r**2)*dV # r^2 dependence.
    return curr


# Define the source point and current vector within the wire
# based on polar, cartesian coordinates and random sampling.

def curr_density_r2_polar (I,R,L,dR,dL):
    dphi = np.pi/32
    r,phi,z = np.meshgrid(np.arange(0,R+dR,dR),
                          np.arange(0,2*np.pi,dphi),
                          np.arange(-L/2,L/2+dL,dL),
                          indexing='ij')
    x, y = r*np.cos(phi), r*np.sin(phi)
    p_curr = np.vstack((x.flatten(),y.flatten(),z.flatten()))
    dV = r.flatten()*dR*dphi
    curr = r2_curr_from_p_curr(I,R,L,p_curr,dV)
    return p_curr, curr

def curr_density_r2_cartesian (I,R,L,dR,dL):
    v = np.arange(-R,R+dR,dR)
    x,y,z = np.meshgrid(v,v,np.arange(-L/2,L/2+dL,dL),
                        indexing='ij')
    p_tmp = np.vstack((x.flatten(),y.flatten(),z.flatten()))
    # So far, we have collected points within a square prism.
    # Just keep the points inside the cylinder.
    r_tmp = np.sqrt(p_tmp[0]**2+p_tmp[1]**2)
    p_curr = p_tmp[:,r_tmp<=R]
    dV = (dR**2)*dL
    curr = r2_curr_from_p_curr(I,R,L,p_curr,dV)
    return p_curr, curr

def curr_density_r2_random (I,R,L,N=10000):
    # We take sqrt to ensure uniform distribution.
    r = np.sqrt(np.random.rand(N))*R
    phi = np.random.rand(N)*2*np.pi
    z = (np.random.rand(N)*2-1)*L/2 # between -L/2 and L/2
    x = r*np.cos(phi)
    y = r*np.sin(phi)
    dV = np.pi*(R**2)*L/N
    p_curr = np.vstack((x.flatten(),y.flatten(),z.flatten()))
```

```
    curr = r2_curr_from_p_curr(I,R,L,p_curr,dV)
    return p_curr, curr
```

```
# Code Block 5.18

# Visualizing the current source points.

def show_r2_density(I,R,L,p_curr,curr,title=''):
    r = np.sqrt(p_curr[0]**2+p_curr[1]**2)
    curr_r = curr[2]
    axis_lim = np.array([-R,R])*1.1

    f,(a0,a1) = plt.subplots(1,2,figsize=(7,3),sharey=True,
                          gridspec_kw={'width_ratios':[1,2]})
    a0.scatter(p_curr[0],p_curr[1],s=1,color='black')
    a0.axis('square')
    a0.set_xlim(axis_lim)
    a0.set_xticks((-R,0,R))
    a0.set_xlabel('x')
    a0.set_ylim(axis_lim)
    a0.set_yticks((-R,0,R))
    a0.set_ylabel('y')

    a1.scatter(p_curr[2],p_curr[1],s=1,color='black')
    a1.set_xlim(np.array([-L/2,L/2])*1.1)
    a1.set_xticks(np.array([-L/2,0,L/2]))
    a1.set_xlabel('z')
    a1.set_ylim(axis_lim)
    a1.set_yticks((-R,0,R))

    f.suptitle('N = %d'%len(r))
    plt.tight_layout()
    plt.savefig('fig_ch5_r2_density_%s.pdf'%title)
    plt.show()


I = 1
R, L = 0.5, 10
dR, dL = R/10, L/30

print('Sampling in the polar coordinates')
p_curr, curr = curr_density_r2_polar (I,R,L,dR,dL)
show_r2_density(I,R,L,p_curr,curr,title='polar')

print('Sampling in the cartesian coordinates')
p_curr, curr = curr_density_r2_cartesian (I,R,L,dR,dL)
show_r2_density(I,R,L,p_curr,curr,title='cartesian')
```

```
print('Sampling randomly')
p_curr, curr = curr_density_r2_random (I,R,L,N=5000)
show_r2_density(I,R,L,p_curr,curr,title='random')
```

Sampling in the polar coordinates

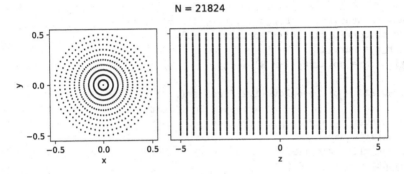

Sampling in the cartesian coordinates

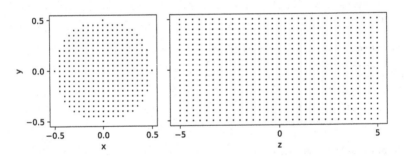

Sampling randomly

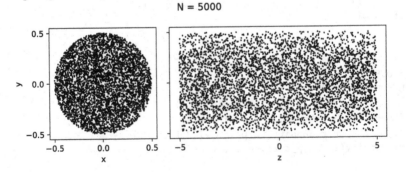

Figure 5.17: Polar, cartesian and random samplings

Let's compare different sampling approaches by calculating the magnetic fields at different distance from the center of the wire. Because of the circular symmetry about the z-axis, we only need to examine one radial direction, such as the positive x-axis. To consider a few points along the x-axis, we set `p_axis = np.arange(0,4*R,R/5)` and assign `p[0] = p_axis` (x coordinates), while `p[1]` and `p[2]` (y and z coordinates) are set to zero.

The following figures illustrate that the result of these calculations may contain some serious artifacts stemming from numerical imprecision and finite sampling. For example, the calculated magnetic field vectors, especially the ones inside the circular cross-section, have erroneous radial components and are not parallel to the y-axis. The strengths of these magnetic field vectors can be grossly over- or under-estimated. These errors are more evident inside of the wire because the observation (p) and the source (p_curr) points can be very closely located. Hence, as discussed in the previous section, the division operation in the `get_magnetic_field()` can amplify the numerical imprecision and produce erroneous results.

```python
# Code Block 5.19

# Plot magnetic fields based on different sampling approaches.

def plot_ball_and_B (R, p_curr, curr, rotate=False, title=''):
    p_axis = np.arange(0,4*R,R/5)
    maxval = np.max(p_axis)
    p = np.zeros((3,len(p_axis))) # observation points
    p[0] = p_axis
    B = get_magnetic_field(p, p_curr, curr)
    B_mag = np.sqrt(np.sum(B**2,axis=0))

    fig, ax = plt.subplots(figsize=(3,3))
    circle = plt.Circle((0,0), R, facecolor ='none')
    ax.add_patch(circle)

    v = np.linspace(-R,R,101)
    x,y = np.meshgrid(v,v)
    current_density = x**2 + y**2
    im = ax.imshow(current_density, clip_path = circle,
                cmap = plt.cm.Greys, extent = [-R, R, -R, R],
                clip_on = True, zorder=0)
    im.set_clip_path(circle)
```

```
    ax.quiver(p[0],p[1],B[0],B[1],angles='xy',scale_units='xy')

    # Show magnetic fields along other directions, too.
    if rotate:
        phi_range = np.arange(np.pi/4,np.pi*2,np.pi/4)
        for phi in phi_range:
            rot_mat = np.array([[np.cos(phi),np.sin(phi)],
                                [-np.sin(phi),np.cos(phi)]])
            rot_px = np.cos(phi)*p[0] - np.sin(phi)*p[1]
            rot_py = np.sin(phi)*p[0] + np.cos(phi)*p[1]
            rot_Bx = np.cos(phi)*B[0] - np.sin(phi)*B[1]
            rot_By = np.sin(phi)*B[0] + np.cos(phi)*B[1]
            ax.quiver(rot_px,rot_py,rot_Bx,rot_By,
                      angles='xy',scale_units='xy')

    N = p_curr.shape[1] # Number of samples
    plt.title('B with $J \sim r^{2}$ (N=%d)'%N)
    plt.axis('square')
    plt.xlabel('x')
    plt.ylabel('y')
    plt.xlim((-maxval,maxval))
    plt.ylim((-maxval,maxval))
    plt.xticks(())
    plt.yticks(())
    plt.tight_layout()
    plt.savefig('fig_ch5_ball_B_%s.pdf'%(title))
    plt.show()
    return p, B_mag

#print('Sampling in the polar coordinates')
p_curr, curr = curr_density_r2_polar (I,R,L,R/20,L/30)
_, _ = plot_ball_and_B (R,p_curr,curr,title='polar')

#print('Sampling in the cartesian coordinates')
p_curr, curr = curr_density_r2_cartesian (I,R,L,R/20,L/30)
_, _ = plot_ball_and_B (R,p_curr,curr,title='cartesian')

#print('Sampling randomly')
p_curr, curr = curr_density_r2_random (I,R,L,N=40000)
_, _ = plot_ball_and_B (R,p_curr,curr,title='random_40000')

# Most accurate (largest N).
# Keep the return values to be used later.
p_curr, curr = curr_density_r2_random (I,R,L,N=10000000)
p_biot, B_mag_biot = plot_ball_and_B (R,p_curr,curr,
                                      title='random_10e7')
```

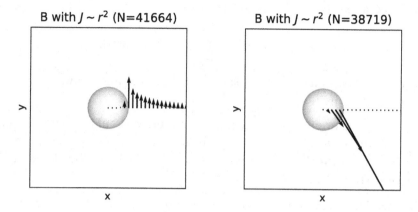

Figure 5.18: Magnetic fields calculated with polar and cartesian samplings.

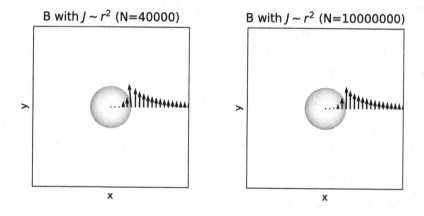

Figure 5.19: Magnetic fields calculated with random sampling.

Among the three different approaches, the random sampling can be most easily scaled with higher N and suffers least from undesired artifacts, because the probability of selecting a random source point very close to any one of the observation points is very small. We can compare our numerical results based on the Biot-Savart law with the exact results based on Ampere's law. Consider a circular loop C with radius r around the cross-section of the wire. Because the magnetic fields point tangentially along the loop (i.e., no radial component), we can calculate the line integral easily, as shown below.

$$\oint_{\text{Contour}} \vec{B} \cdot \vec{dl} = B \oint_{\text{Contour}} ds = 2\pi r B,$$

where B is the strength of the magnetic field at distance r away from the center of the wire.

According to Ampere's law, the above quantity should be equal to μ_0 times the amount of current enclosed by the Amperian loop. When $r > R$, it is easy because the total amount of current enclosed by this loop is always I. Hence, $B = \frac{\mu_0 I}{2\pi r}$.

The case for $r < R$ is a bit more complicated since the total amount of enclosed current depends on r.

$$I_{\text{enclosed}} = \int_0^r \int_0^{2\pi} \vec{J} \cdot \hat{n} r' dr' d\phi = \frac{\alpha \pi}{2} r^4.$$

Hence, the magnetic field inside the wire is $B = \frac{\mu_0 I_{\text{enclosed}}}{2\pi r} = \frac{\mu_0 \alpha}{4} r^3$.

The following code block plots the magnetic field strengths computed with Ampere's and Biot-Savart laws.

```
# Code Block 5.20

# Comparison of Biot-Savart and Ampere's laws.

mu0 = 4*np.pi*(10**-7)
B_norm = mu0*I/(2*np.pi*R) # Maximum B with the known I and R.

# Ampere's law with J ~ r**2
alpha = (2*I)/(np.pi*(R**4))
r_axis = np.arange(0,4*R,R/5)
r = np.zeros((3,len(r_axis)))
r[0] = r_axis
B_ampere = np.zeros((3,len(r_axis)))

# Direction of the field can be inferred from right-hand rule,
# which is +y-direction in our example.
B_ampere[1][r_axis<=R] = mu0*alpha*(r[0][r_axis<=R]**3)/4
B_ampere[1][r_axis>=R] = mu0*I/(2*np.pi*r[0][r_axis>=R])

fig, ax = plt.subplots(figsize=(3,3))
circle = plt.Circle((0,0), R, facecolor ='none')
ax.add_patch(circle)

v = np.linspace(-R,R,101)
```

```
x,y = np.meshgrid(v,v)
current_density = x**2 + y**2
im = ax.imshow(current_density, clip_path = circle,
               cmap = plt.cm.Greys, extent = [-R, R, -R, R],
               clip_on = True, zorder=0)
im.set_clip_path(circle)
ax.quiver(r[0],r[1],B_ampere[0],B_ampere[1],
          angles='xy',scale_units='xy')

plt.title("B from Ampere's law")
plt.axis('square')
plt.xlabel('x')
plt.ylabel('y')
plt.xlim((-np.max(r_axis),np.max(r_axis)))
plt.ylim((-np.max(r_axis),np.max(r_axis)))
plt.xticks(())
plt.yticks(())
plt.tight_layout()
plt.savefig('fig_ch5_ball_B_ampere.pdf')
plt.show()

# Comparing the magnitudes of B from Biot-Savart and Ampere.
plt.figure(figsize=(6,3))
plt.scatter(p_biot[0]/R, B_mag_biot/B_norm, s=50,
            color='gray', label ="Biot-Savart")
plt.plot(r[0]/R,B_ampere[1]/B_norm,
         color='black', label="Ampere")

plt.title("Comparison of Biot-Savart and Ampere")
plt.xticks((0,1,2,3,4))
plt.xlabel('$r/R$')
plt.ylabel('normalized B')
plt.ylim((-0.1,1.1))
plt.yticks((0,0.5,1))
plt.legend(framealpha=1)
plt.savefig('fig_ch5_ball_B_biot_ampere.pdf')
plt.show()
```

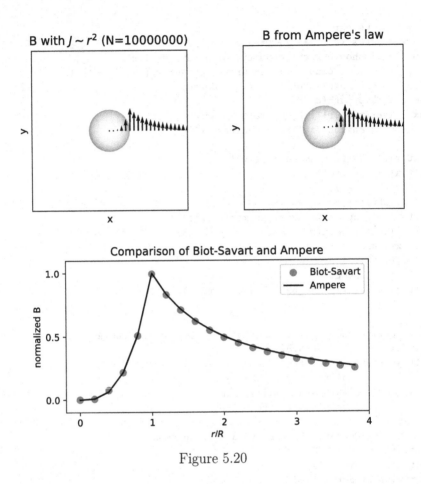

Figure 5.20

Considering the numerical imperfection seen while working with the Biot-Savart law in the above example, one might ask if we should always be using Ampere's law to calculate the magnetic fields. Unfortunately, Ampere's law is effective only when there is an exploitable geometrical symmetry. In the above example, a circular symmetry allowed us to easily calculate the line integral in Ampere's law, where we used the fact that B is a constant along a circular contour. When there is no symmetry in the current distribution, we have to rely on the Biot-Savart law.

5.5 CURL

Imagine a windmill under three different wind conditions, as shown in the following figure. In all cases, the wind is blowing from left to right.

In the first case, the wind is stronger at the top, so the windmill rotates in a clockwise direction. The wind is stronger at the bottom in the third case, and it rotates in a counterclockwise direction. If the wind is blowing uniformly, as in the second case, the windmill will not rotate because the net torque is zero.

```python
# Code Block 5.21

# Visualize the circulation of a vector field.

# To determine the circulation, we need a vector field definition
# and a closed path around which we can calculate the dot product
# between its line segments and the vectors.

# Define a closed path as a circle centered at the origin.
path, _ = currents_along_circle (I=0, R=0.5, d_phi=np.pi/128)
x, y = np.meshgrid(np.arange(-1,1,0.5),np.arange(-1,1.5,0.5),
                   indexing='ij')
p = np.vstack((x.flatten(),y.flatten()))

def get_vfield_example (p,case=1):
    strength = 100/2
    vfield = np.zeros(p.shape)
    if case==0: # Top-heavy wind
        vfield[0] = strength*(1+p[1])
    if case==1: # Equal wind
        vfield[0] = strength*2
    if case==2: # Bottom-heavy wind
        vfield[0] = strength*(1-p[1])
    return vfield

def plot_full_field (ax,p,vfield,scale=250):
    ax.quiver(p[0],p[1],vfield[0],vfield[1],
              angles='xy',scale_units='xy',scale=scale)
    return

def plot_around_path (ax,path,vfield,scale=250):
    s = 16
    ax.plot(path[0],path[1],color='gray',linewidth=1)
    ax.quiver(path[0][::s],path[1][::s],
              vfield[0][::s],vfield[1][::s],
              angles='xy',scale_units='xy',scale=scale)
    return

def plot_windmill (ax,rotate_mag=0):
    R = 1.0
    c = ('#EEEEEE','#CCCCCC','#AAAAAA','#000000')
    phi_range = np.array([0,1,2,3])*np.pi/20*rotate_mag
```

```
    for i, phi in enumerate(phi_range):
        # Plot blades of the windmill.
        ax.plot([-R*np.cos(-phi),+R*np.cos(-phi)],
                [-R*np.sin(-phi),+R*np.sin(-phi)],
                color=c[i],linewidth=6)
        ax.plot([-R*np.cos(-phi+np.pi/2),+R*np.cos(-phi+np.pi/2)],
                [-R*np.sin(-phi+np.pi/2),+R*np.sin(-phi+np.pi/2)],
                color=c[i],linewidth=6)
    return

def tidy_axis(ax):
    lim = 1.25
    ax.axis('square')
    ax.set_xticks(())
    ax.set_yticks(())
    ax.set_xlim((-lim,lim))
    ax.set_ylim((-lim,lim))

scale = 250
fig, axs = plt.subplots(3,3,figsize=(5,5),sharey=True,sharex=True)

for i in range(3):
    full_vfield = get_vfield_example(p,case=i)
    vfield = get_vfield_example(path,case=i)
    circulation = line_integral_vector_field(vfield,path)

    plot_full_field (axs[0,i],p,full_vfield,scale=scale)
    tidy_axis(axs[0,i])

    plot_windmill(axs[1,i],rotate_mag=circulation)
    tidy_axis(axs[1,i])

    plot_around_path(axs[2,i],path,vfield,scale=scale)
    tidy_axis(axs[2,i])
    axs[2,i].set_title("%3.2f"%circulation)

axs[0,0].set_title('Top-heavy')
axs[0,1].set_title('Uniform')
axs[0,2].set_title('Bottom-heavy')
axs[1,0].set_title('Clockwise')
axs[1,1].set_title('No Rotation')
axs[1,2].set_title('Counterclockwise')
axs[0,0].set_ylabel('Full Field')
axs[1,0].set_ylabel('Windmill')
axs[2,0].set_ylabel('Path')

plt.tight_layout()
plt.savefig('fig_ch5_wind_circulation.pdf',bbox_inches='tight')
plt.show()
```

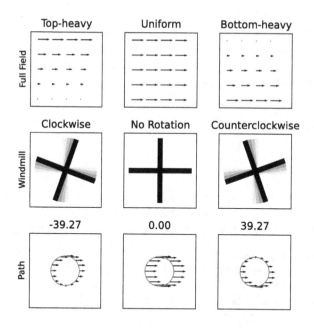

Figure 5.21

For a given vector field $\vec{K}$ like a wind map, we define a mathematical quantity called circulation as a line integral around a closed path: $\oint_{\text{Contour}} \vec{K} \cdot \vec{dl}$. The sign of this integral indicates the direction of the rotation, and its magnitude is proportional to the degree of "swirling" of a vector field.

A related mathematical operation is the curl of a vector field. As discussed in Chapter 2, it is defined as:

$$\nabla \times \vec{K} = \left(\frac{\partial K_z}{\partial y} - \frac{\partial K_y}{\partial z}\right)\hat{x} + \left(\frac{\partial K_x}{\partial z} - \frac{\partial K_z}{\partial x}\right)\hat{y} + \left(\frac{\partial K_y}{\partial x} - \frac{\partial K_x}{\partial y}\right)\hat{z}.$$

For example, for a vector field $\vec{K}(x, y, z) = (x^2 + y)\hat{x} + (y + z^2)\hat{y} + (x + y)\hat{z}$, its curl is: $\nabla \times \vec{K} = (1 - 2z)\hat{x} - \hat{y} - \hat{z}$. The following code block verifies this using symbolic mathematics with sympy. Here, the components of the curl vector are defined with the first derivatives of the vector field, obtained with the function sym.diff().

```
# Code Block 5.22

# curl with sympy

x, y, z = sym.symbols('x y z')

# vector field example
vx = x**2+y
vy = y+z**2
vz = x+y

# Calculation of curl
curlx=(sym.diff(vz,y)-sym.diff(vy,z))
curly=(sym.diff(vx,z)-sym.diff(vz,x))
curlz=(sym.diff(vy,x)-sym.diff(vx,y))

print("\tVector Field:",[vx,vy,vz])
print("\tCurl:",[curlx,curly,curlz])
```

```
        Vector Field: [x**2 + y, y + z**2, x + y]
        Curl: [1 - 2*z, -1, -1]
```

Curl is defined at each point of a vector field. The following code block shows examples of a vector field and their respective curls evaluated at various points.

```
# Code Block 5.23

# Visualization of curl

# Variables for symbolic calculation.
x,y,z = sym.symbols('x y z')

# Variables for numerical calculation.
step = 0.25
v = np.arange(-1,1+step,step)
x_range,y_range,z_range = np.meshgrid(v,v,0,indexing='ij')
p = np.vstack((x_range.flatten(),y_range.flatten(),z_range.flatten()))
N = p.shape[1]

p_line, _ = currents_along_square (I=0, L=2, dL=step)

for s in range(2):
    print("Example",s+1)
    if s==0:
        vx,vy,vz = -y,x,sym.Integer(0)
        v_line = np.vstack(((-p_line[1],p_line[0])))
    if s==1:
        vx,vy,vz = -y**2,x**2,sym.Integer(0)
```

```
        v_line = np.vstack((-p_line[1]**2,p_line[0]**2))

curlx = sym.diff(vz,y)-sym.diff(vy,z)
curly = sym.diff(vx,z)-sym.diff(vz,x)
curlz = sym.diff(vy,x)-sym.diff(vx,y)

print("\tVector Field:",[vx,vy,vz])
print("\tCurl:",[curlx,curly,curlz])

v = np.zeros((3,N))
curl = np.zeros((3,N))
for i in range(N):
    v[0,i] = vx.subs([(x,p[0][i]),(y,p[1][i]),(z,p[2][i])])
    v[1,i] = vy.subs([(x,p[0][i]),(y,p[1][i]),(z,p[2][i])])
    v[2,i] = vz.subs([(x,p[0][i]),(y,p[1][i]),(z,p[2][i])])
    curl[0,i] = curlx.subs([(x,p[0][i]),(y,p[1][i]),(z,p[2][i])])
    curl[1,i] = curly.subs([(x,p[0][i]),(y,p[1][i]),(z,p[2][i])])
    curl[2,i] = curlz.subs([(x,p[0][i]),(y,p[1][i]),(z,p[2][i])])

fig = plt.figure(figsize=(6,3))
grid = fig.add_gridspec(1,2)

ax0 = fig.add_subplot(grid[0,0])
ax0.quiver(p[0],p[1],v[0],v[1], scale=5, color='gray')
ax0.axis('square')
lim = 2
ax0.set_xlim(np.array([-1,1])*lim)
ax0.set_ylim(np.array([-1,1])*lim)
ax0.set_xticks((-1,0,1))
ax0.set_yticks((-1,0,1))
ax0.set_xlabel('x')
ax0.set_ylabel('y')
ax0.set_title('Vector Field')

ax1 = fig.add_subplot(grid[0,1], projection='3d')
ax1.quiver(p[0],p[1],p[2],curl[0],curl[1],curl[2],
           length=0.2,color='black')
ax1.set_zticks((-1,0,1))
ax1.set_zlim(np.array([-1,1]))
lim = 1.25
ax1.set_xlim(np.array([-1,1])*lim)
ax1.set_ylim(np.array([-1,1])*lim)
ax1.set_xticks((-1,0,1))
ax1.set_yticks((-1,0,1))
ax1.set_xlabel('x')
ax1.set_ylabel('y')
ax1.set_zlabel('z')
ax1.set_title('Curl')
```

```
plt.tight_layout()
plt.savefig('fig_ch5_curl_ex%d.pdf'%(s+1),bbox_inches='tight')
plt.show()
```

Example 1
        Vector Field: [-y, x, 0]
        Curl: [0, 0, 2]

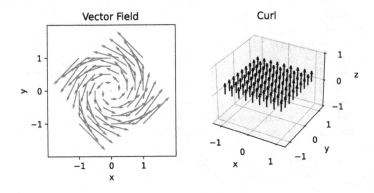

Figure 5.22

Example 2
        Vector Field: [-y**2, x**2, 0]
        Curl: [0, 0, 2*x + 2*y]

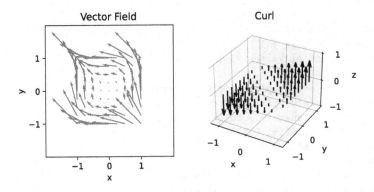

Figure 5.23

The first example shows a classic vector field with a vortex-like swirling around the z-axis. In this case, $\vec{K} = -y\hat{x} + x\hat{y}$ and $\nabla \times \vec{K} = 2\hat{z}$. The convention of the right-hand rule, which defines the relative orientations of x, y, and z axes, applies here. As we follow the swirling of the vector field with our four right fingers, the right thumb points in the direction of the rotation axis. This is the same convention used in classical mechanics for defining an object's rotation or angular momentum. The curl vector calculated at each point of the vector field consistently points along the $+z$-axis in this example.

The second example shows a more complex vector field, where the swirling of the vector field is counterclockwise in the upper-right corner (first quadrant) of the xy-plane, while the lower-left corner (third quadrant) shows clockwise swirling. It is also apparent that the vector field's swirling increases as we move away from the origin along the diagonal direction of $y = x$. The curl of this vector field is, therefore, positive in the first quadrant and negative in the third quadrant. The magnitude of the curl vectors increases as we move away from the origin.

According to the curl theorem, also known as Stoke's law, the sum of curls of a vector field $\vec{\mathbf{K}}$ over a region of interest is equal to the total circulation of the vector field calculated along the region's boundary.

$$\iint_{\text{Area}} (\nabla \times \vec{\mathbf{K}}) \cdot d\vec{a} = \oint_{\text{Contour}} \vec{\mathbf{K}} \cdot d\vec{l}.$$

The differential areal element $d\vec{a}$ is defined on the region of interest, which is enclosed by the contour boundary. The differential line segment $d\vec{l}$ is defined along the one-dimensional closed contour.

The following code block compares the areal or surface integral of the curls (left side of Stokes' theorem) and the line integral of the vector field (right side of Stokes' theorem). We use the same vector field examples from the above, evaluate the integrals with different boundary shapes and locations, and demonstrate that these two methods of calculations yield the same value within the numerical precision of the step parameter. For the surface integral, we will use the surface area on the xy-plane.

In the first vector field example, where the curls are uniform with $\nabla \times \vec{\mathbf{K}} = 2\hat{z}$, the surface integral can be trivially calculated as $2A$, where A is the area of the region. Hence, the numerical results can be easily checked against the exact value. The analytical calculation is not as trivial for the second example. However, we note that when the boundary is centered at the origin or at x0, y0 = 0.5,-0.5, the symmetry of this particular vector field forces the line integral or the surface integral to be exactly equal to zero. The numerical results are consistent with this observation.

```python
# Code Block 5.24

# Demonstration of Stokes' theorem with square loops.

def demo_stokes (p_boundary, step=0.001, skip=200, title=''):

    # Variables for numerical calculation.
    dxdy = step**2
    fig, axs = plt.subplots(2,3,figsize=(7,6),
                            sharey=True,sharex=True)

    for s in range(2):
        # Use the same examples as before.
        # curl has no x or y components, so calculation is simple.
        for ctr in range(3):

            x,y = np.meshgrid(np.arange(-1/2,1/2,step),
                              np.arange(-1/2,1/2,step),
                              indexing='ij')
            p_area = np.vstack((x.flatten(),y.flatten()))

            # Special consideration for the circular area.
            if title=='circle':
                r = np.sqrt(p_area[0]**2+p_area[1]**2)
                # Estimate R, assuming a circle at the origin.
                R = np.max(p_boundary[0])
                p_area = p_area[:,r<=R]

            # Different boundary cases
            if ctr==0:
                x0,y0 = 0,0
            if ctr==1:
                x0,y0 = -0.25, -0.5
            if ctr==2:
                x0,y0 = 0.5, -0.5

            p_line = np.zeros(p_boundary.shape)
            p_line[0], p_line[1] = p_boundary[0]+x0, p_boundary[1]+y0
            p_area[0], p_area[1] = p_area[0]+x0, p_area[1]+y0

            # Different examples
            if s==0:
                v_line = np.vstack((-p_line[1],p_line[0]))
                curlz = 2+np.zeros(p_area.shape[1])
            if s==1:
                v_line = np.vstack((-p_line[1]**2,p_line[0]**2))
```

```
              curlz = 2*(p_area[0]+p_area[1])

          method1 = line_integral_vector_field (v_line,p_line)
          method2 = np.sum(curlz)*dxdy

          ax = axs[s,ctr]
          ax.plot(p_line[0],p_line[1],color='black',linewidth=1)
          ax.quiver(p_line[0][::skip],p_line[1][::skip],
                    v_line[0][::skip],v_line[1][::skip],
                    scale=4,color='gray')
          del p_line, p_area

          ax.axis('square')
          lim = 2
          ax.set_xlim(np.array([-1,2])*lim)
          ax.set_ylim(np.array([-1,2])*lim)
          ax.set_xticks((-1,0,1,2))
          ax.set_yticks((-1,0,1,2))
          ax.set_xlabel('x')
          ax.set_ylabel('y')
          ax.set_title('Example %d'%(s+1))
          ax.legend((r'$\oint_C$: %+4.3f'%method1,
                     r'$\iint_A$: %+4.3f'%method2),
                    handlelength=0)

   plt.tight_layout()
   plt.savefig('fig_ch5_stokes_%s.pdf'%(title),bbox_inches='tight')
   plt.show()
```

```
# Code Block 5.25
step = 0.001

p_boundary, _ = currents_along_square (I=0,L=1,dL=step)
demo_stokes(p_boundary,title='square')

p_boundary, _ = currents_along_circle (I=0,R=0.5,d_phi=step)
demo_stokes(p_boundary,title='circle',skip=250)
```

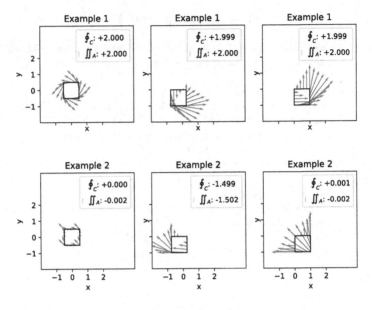

Figure 5.24

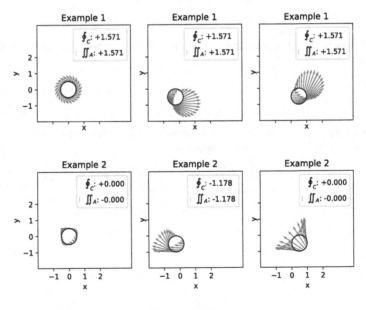

Figure 5.25

It is also worth mentioning that the curl theorem has a similar surface independence as the path independence of potential energy change under the conservative force. As long as the closed contour boundary is the same, the surface on which we perform the two-dimensional areal integral does not matter. There are infinitely many surfaces enclosed by the same contour boundary. The following figures show three different surfaces defined by the same circular boundary of $x^2 + y^2 = R^2$ with $R = 0.5$. Any of them can be used to calculate the surface integral of the curl theorem, and they will yield identical results. The flat surface on the xy-plane (first figure) would, of course, be an obvious choice for calculating the integral because of its simplicity. The normal vectors on the other surfaces would vary depending on their positions, whereas the flat surface has the same normal vector of $+\hat{z}$ at every point.

Similar to our earlier discussion about the flux of a vector field through a surface, the direction of the normal vector $d\vec{a}$ of the surface must be specified. Conventionally, it is determined by the right-hand rule. As you wrap the four fingers of your right hand along the closed path (again, the positive direction will be a counterclockwise direction when we view the xy-plane from the positive z-axis), the heading of the thumb will be the general direction of the normal vector. We can express the surface integral of the curl theorem as $\iint_{\text{Area}} (\nabla \times \vec{\mathbf{K}}) \cdot \hat{n} da$. We will not prove Stokes' theorem in this book but encourage the readers to consult other resources on vector calculus.

```
# Code Block 5.26

# Visualize different surfaces with the same boundary.

R = 0.5
dR = 0.01
r, phi, z = np.meshgrid(np.arange(0,R,dR),
                        np.arange(0,2*np.pi,np.pi/100),
                        0, indexing='ij')
x, y = r*np.cos(phi), r*np.sin(phi)

# Same boundary.
p = np.vstack((x.flatten(),y.flatten(),z.flatten()))

# Different examples.
surface = [0*(p[0]+p[1]),
           -np.cos((3*np.pi/2)*(p[0]**2+p[1]**2)/(0.5**2)),
           (1.5*np.exp(-((p[0]**2+p[1]**2)-0.5**2)/0.3))-1.5]
```

```
print('These surfaces have the same boundary.')

fig = plt.figure(figsize=(8,3))
grid = fig.add_gridspec(1,3)
for i in range(3):
    ax = fig.add_subplot(grid[0,i], projection='3d')
    ax.plot_trisurf(p[0],p[1],surface[i],cmap='gray')
    ax.set_xticks(())
    ax.set_yticks(())
    ax.set_zticks(())
    ax.set_xlabel('')
    ax.set_ylabel('')
    ax.set_zlabel('')
    ax.set_xlim((-0.55,0.55))
    ax.set_ylim((-0.55,0.55))
    ax.set_zlim((-1,2))

plt.tight_layout()
plt.savefig('fig_ch5_diff_surfaces_same_boundary.pdf')
plt.show()
```

```
These surfaces have the same boundary.
```

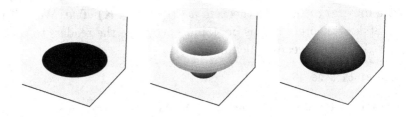

Figure 5.26

5.6 AMPERE'S LAW, AGAIN

Using the curl theorem, let us examine Ampere's law again. We can replace $\vec{K}$ with $\vec{B}$, so that

$$\oint_{\text{Contour}} \vec{B} \cdot \vec{dl} = \iint_{\text{Area}} (\nabla \times \vec{B}) \cdot \hat{n} da.$$

According to Ampere's law,

$$\oint_{\text{Contour}} \vec{B} \cdot \vec{dl} = \mu_0 \sum I = \mu_0 \iint_{\text{Area}} \vec{J} \cdot \hat{n} da.$$

This situation is reminiscent of Gauss's law, which states that the amount of the electric field flux is independent of the shape of the surface of the integral as long as the surface encloses the same amount of electrical charges. The line integral of a magnetic field is independent of the shape of the closed contour (and the surface defined by it) as long as the contour encloses the same amount of current.

The integrand of the right-hand sides must be equal to each other since the area can be an arbitrary surface for any given contour boundary.

$$\nabla \times \vec{B} = \mu_0 \vec{J}.$$

This is the differential form of Ampere's law.

## CHAPTER 6

Force

6.1 ELECTRIC FORCE

Wind moves clouds and air balloons. Meteorologists use wind maps to analyze the weather pattern and to make a forecast. Planets, suns, and galaxies exert gravitational force on each other. Astronomers can analyze their interactions and predict the movement of astronomical objects. Likewise, electrical charges exert electric force on one another via electric fields, influencing the charges' motion. More precisely, the electric force $\vec{\mathbf{F}}_{\text{electric}}$ on a charge q located at a location is proportional to the electric field $\vec{\mathbf{E}}$ at that exact position.

$$\vec{\mathbf{F}}_{\text{electric}} = q\vec{\mathbf{E}}.$$

This suggests that the electric field is the force on a unit charge. Consider an electric charge Q placed at the origin that sets up an electric field in three dimensions:

$$\vec{\mathbf{E}}(x, y, z) = \frac{1}{4\pi\epsilon_0} \frac{Q}{r^2} \hat{\mathbf{r}},$$

where $r^2 = x^2 + y^2 + z^2$. Another charge q located at position (x, y, z) will thus experience the electric force by Q's electric field:

$$\vec{\mathbf{F}}_{\text{electric}} = \frac{1}{4\pi\epsilon_0} \frac{qQ}{r^2} \hat{\mathbf{r}}.$$

DOI: 10.1201/9781003397496-6

The above expression is called the Coulomb's law of electric force. Note that this is symmetric for q and Q, and it can be thought of as the electric force exerted on Q by the electric field of q. The line of reasoning that involves an electric field addresses the question of action at a distance. How do two electrical charges separated by a distance exert force on each other? The answer is that they interact through their electric fields.

This single formulation captures different sign combinations of electric charges and the direction of the force. If q and Q are both positive, the resulting force has a positive sign, and q will be pushed away from Q. If q and Q are both negative, the resulting force again takes on a positive sign, indicating that the force is again repulsive. If Q and q have the opposite signs, the force is then attractive. The respective forces on q and Q are equal in magnitude and opposite in direction.

6.2   MAGNETIC FORCE

A charge q moving with velocity $\vec{v}$ experiences the magnetic force due to magnetic field $\vec{B}$ at that position.

$$\vec{F}_{magnetic} = q\vec{v} \times \vec{B}.$$

Unlike the electric force, which lines up with the direction of the electric field, the magnetic force is orthogonal to both $\vec{v}$ and $\vec{B}$. The sign of q influences the direction of the force. The magnitude of the force depends on the amount of charge, the magnitude of the velocity, the strength of the magnetic field, and the angle between $\vec{v}$ and $\vec{B}$.

Why is the magnetic force so much more complicated and "unintuitive" than electric? We will have to accept it as it is, since it is verified and confirmed by numerous rigorous experiments. We will also learn in the later chapters (on Faraday's law and Einstein's special theory of relativity) that this form of magnetic force makes electromagnetism wondrously unified and consistent.

Consider a region of space with a uniform magnetic field, which may be set up with a solenoid. The following figure illustrates this uniform magnetic field with regularly spaced dots. Each dot, representing a head-on view of an arrow, indicates that the field is pointing out of a page. In order to indicate the field that points into a page, we use a marker x, which looks like the rear end of an arrow. The velocity of the charge is indicated with a gray arrow, and the resulting magnetic force is marked with a black arrow. The standard units of the velocity, force, and magnetic field are m/s, N, and T, respectively.

```python
# Code Block 6.1

# Relationship between magnetic force, velocity, and field.

import numpy as np
import matplotlib.pyplot as plt

# Define the amount of charge, velocity, position,
# and the magnetic field.
q = 2
x, y = -1, -1.5
vx, vy, vz = -1, 1, 0 # vz=0, so it moves on the xy plane.
Bx, By, Bz = 0, 0, 1 # Bx=By=0, so B field is along z.

# magnetic force F = q*(v x B)
Fx, Fy = q*(vy*Bz-vz*By), q*(vz*Bx-vx*Bz)

fig = plt.figure(figsize=(3,3))
plt.quiver(x,y,vx,vy,color='gray',
            angles='xy',scale_units='xy',scale=1)
plt.quiver(x,y,Fx,Fy,color='black',
            angles='xy',scale_units='xy',scale=1)

lim, step = 3, 1
xgrid, ygrid = np.meshgrid(np.arange(-lim,lim+step,step),
                            np.arange(-lim,lim+step,step),
                            indexing='ij')

plt.scatter(xgrid,ygrid,marker='.',color='gray')
plt.scatter(x,y,s=100,marker='o',color='black')
plt.legend(('velocity [m/s]','Force [N]','B field [T]'),
            framealpha=1)
plt.axis('square')
plt.xlim((-lim-0.5,lim+0.5))
plt.ylim((-lim-0.5,lim+0.5))
plt.xticks((-lim,0,lim))
plt.yticks((-lim,0,lim))
plt.xlabel('x')
plt.ylabel('y')
plt.tight_layout()
plt.savefig('fig_ch6_mag_force.pdf')
plt.show()
```

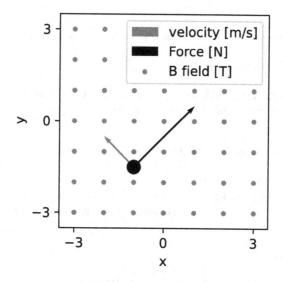

Figure 6.1

Because the magnetic force is always orthogonal to the velocity of an electrically charged particle, there is no work associated with the magnetic force. The acceleration caused by the magnetic force will change the direction of a moving charge, but it will not change its speed. An interesting application of this is to trap a charged particle with mass m within a magnetic field with centripetal acceleration $a_c = \frac{v^2}{R}$ for uniform circular motion with radius R.

$$a_c = \frac{v^2}{R} = \frac{F_{\text{magnetic}}}{m} = \frac{qvB}{m},$$

which implies that $R = \frac{mv}{qB}$. A mass spectrometer operates based on this principle that R is related to m. If the amount of charge q and speed v of a molecule are known, its mass can be determined by measuring how much the molecule bends its trajectory inside a uniform magnetic field B. The following code block illustrates a uniform circular motion of a trapped charged particle.

```
# Code Block 6.2

# a charge trapped in a circular orbit under uniform B field.

fig = plt.figure(figsize=(5,5))

# Define the charge, velocity, and position of a charge
phi_range = np.arange(0,2*np.pi,2*np.pi/7)
v = -1
q = 1
R = 1.75 # This effectively fixes m = qBR/v.

lim, step = 3, 1
xgrid, ygrid = np.meshgrid(np.arange(-lim,lim+step,step),
                           np.arange(-lim,lim+step,step),
                           indexing='ij')

for phi in phi_range:
    x, y = R*np.cos(phi), R*np.sin(phi)
    vx, vy, vz = -np.sin(phi)*v, np.cos(phi)*v, 0
    Bx, By, Bz = 0, 0, 1 # Bx=By=0, so B field is along z.
    Fx, Fy = q*(vy*Bz-vz*By), q*(vz*Bx-vx*Bz) # F = q*(v x B)

    plt.quiver(x,y,vx,vy,color='gray',
               angles='xy',scale_units='xy',scale=1)
    plt.quiver(x,y,Fx,Fy,color='black',
               angles='xy',scale_units='xy',scale=1)
    plt.scatter(xgrid,ygrid,marker='.',color='gray')
    plt.scatter(x,y,s=100,marker='o',color='black')

plt.legend(('velocity [m/s]','Force [N]','B field [T]'),
           framealpha=1)
plt.axis('square')
plt.xlim((-lim-0.5,lim+0.5))
plt.ylim((-lim-0.5,lim+0.5))
plt.xticks((-lim,0,lim))
plt.yticks((-lim,0,lim))
plt.xlabel('x')
plt.ylabel('y')
plt.tight_layout()
plt.savefig('fig_ch6_mag_circ.pdf')
plt.show()
```

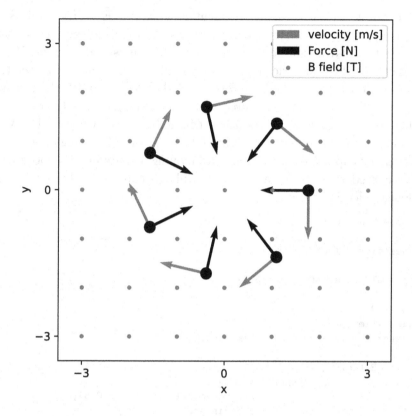

Figure 6.2

6.3 LORENTZ FORCE

When both electric and magnetic fields are present, the full electromagnetic force exerted on a charge q is given by

$$\vec{F} = q\vec{E} + q\vec{v} \times \vec{B},$$

which is known as the Lorentz force.

We can engineer a region of space with constant electric field $\vec{E}$ (with a pair of oppositely charged parallel plates: a capacitor) and constant magnetic field $\vec{B}$ (with loops of steady currents: a solenoid). Furthermore, we can arrange the directions of these two fields so that the electric and magnetic forces on a moving charge are in opposite directions.

An example of such an arrangement is $\vec{v} = v\hat{x}$, $\vec{E} = E\hat{y}$, and $\vec{B} = B\hat{z}$, where v, E, and B are positive values. You can quickly verify that the electric and magnetic forces point in the opposite directions. If the magnitudes of the electric and magnetic fields are controlled to match the strength of these two opposite forces (or $q\vec{E} + q\vec{v} \times \vec{B} = 0$), the moving charge will experience zero force and zero acceleration. This region is called the velocity selector since a charged particle whose velocity is equal to $v = E/B$ can move straight and emerge out of this region without hitting the capacitor plates. Particles moving at a different velocity will have curved trajectories and not be able to come straight out of the velocity selection region.

```python
# Code Block 6.3

# velocity selector

fig = plt.figure(figsize=(6,3))

step = 2
xlim, ylim = 4, 1.5
x_range = np.arange(-xlim,xlim+step,step)

# Draw B field
xgrid, ygrid = np.meshgrid(x_range,
                           np.arange(-ylim,ylim+0.5,0.5),
                           indexing='ij')
plt.scatter(xgrid,ygrid,marker='.',color='gray',label='B')

# Draw charged parallel plates that produces E field.
xgrid_top, ygrid_top = np.meshgrid(x_range,+2.0,indexing='ij')
xgrid_bot, ygrid_bot = np.meshgrid(x_range,-2.2,indexing='ij')
plt.scatter(xgrid_top,ygrid_top,marker='_',s=50,color='black')
plt.scatter(xgrid_bot,ygrid_bot,marker='+',s=50,color='black')
plt.plot([-xlim,xlim],[+ylim,+ylim],color='black',linewidth=7)
plt.plot([-xlim,xlim],[-ylim,-ylim],color='black',linewidth=7)

# Draw a charge and the force on it.
x, y = -1, 0
v = 3
F = 1
plt.quiver(x,y,0,+F,angles='xy',scale_units='xy',scale=1,
           linewidth=2,label='F (electric)',color='black')
plt.quiver(x,y,0,-F,angles='xy',scale_units='xy',scale=1,
           linewidth=2,label='F (magnetic)',color='gray')
plt.quiver(x,y,+v,0,angles='xy',scale_units='xy',scale=1,
           label='v = E/B',color='#CCCCCC')
```

```
plt.scatter(x,y,s=100,marker='o',color='black')

plt.quiver(-10,y,+v,0,color='#CCCCCC',
          angles='xy',scale_units='xy',scale=1)
plt.scatter(-10,y,s=100,marker='o',color='black')

plt.quiver(9,y,+v,0,color='#CCCCCC',
          angles='xy',scale_units='xy',scale=1)
plt.scatter(9,y,s=100,marker='o',color='black')

plt.xlim((-13,15))
plt.ylim((-3,5))
plt.xticks(())
plt.yticks(())
plt.legend()
plt.tight_layout()
plt.savefig('fig_ch6_vel_selector.pdf')
plt.show()
```

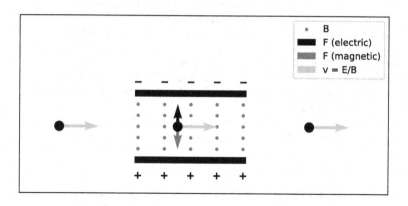

Figure 6.3

6.4 THOMSON'S EXPERIMENT

In the late 1800s, a British physicist, J.J. Thomson, used a cathode ray tube to create a steady beam of electrons and passed them through a velocity selector. By controlling the magnitudes of the electric and magnetic fields and making the beam move straight, he determined the velocity of the electrons as $v_x = E/B$.

When the electric field is turned off, the electron beam would no longer have a straight trajectory. Without the cancelation of the electromagnetic forces, the electrons would gain velocity in the y-direction due to the remaining magnetic force $F_y = qv_xB$, in addition to the initial horizontal velocity of v_x. The final velocity is given by

$$\vec{\mathbf{v}}_f = v_x\hat{\mathbf{x}} + v_y\hat{\mathbf{y}}.$$

```python
# Code Block 6.4

# Deflection of an electron beam with E field off.

fig = plt.figure(figsize=(6,3))

step = 2
xlim, ylim = 4, 1.5

# Draw B.
xgrid, ygrid = np.meshgrid(np.arange(-xlim,xlim+step,step),
                           np.arange(-ylim,ylim+0.5,0.5),
                           indexing='ij')
plt.scatter(xgrid,ygrid,marker='.',color='gray',label='B')

# Draw parallel plates (without charges)
plt.plot([-xlim,xlim],[-ylim,-ylim],color='black',linewidth=7)
plt.plot([-xlim,xlim],[+ylim,+ylim],color='black',linewidth=7)
plt.text(0,ylim+0.5,'L')

# Draw a charge and the force on it.
x, y = -1, 0
v = 3
F = 1
plt.quiver(x,y,0,-F,angles='xy',scale_units='xy',scale=1,
           linewidth=2,label='F (magnetic)',color='gray')
plt.quiver(x,y,+v,-0.2,angles='xy',scale_units='xy',scale=1,
           label='v',color='#CCCCCC')
plt.scatter(x,y,s=100,marker='o',color='black')

plt.quiver(-10,y,+v,0,color='#CCCCCC',
           angles='xy',scale_units='xy',scale=1)
plt.scatter(-10,y,s=100,marker='o',color='black')
plt.text(-10,y+0.5,r'$v = E/B$')

y = -1
plt.quiver(9,y,+v,-0.75,color='#CCCCCC',
           angles='xy',scale_units='xy',scale=1)
plt.scatter(9,y,s=100,marker='o',color='black')
plt.text(10,y,r'$v_f$')
```

```
plt.xlim((-13,15))
plt.ylim((-3,5))
plt.xticks(())
plt.yticks(())
plt.legend()
plt.tight_layout()
plt.savefig('fig_ch6_thomson.pdf')
plt.show()
```

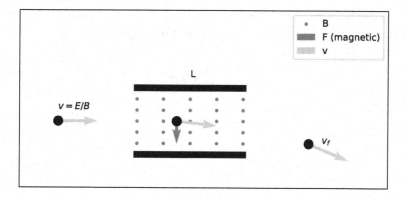

Figure 6.4

As a first-order approximation, assume that the horizontal extent of the velocity selector L is short enough, so the electrons spend very little time inside the region and their paths do not bend too much. Then, the motion inside the velocity selector region is like an elementary kinematics problem where a projectile is shot horizontally and undergoes uniform vertical acceleration due to gravity. What is the vertical velocity after traveling a distance of L? This answer can be derived as follows. The time an electron spends traversing the distance L is $t = \frac{L}{v_x}$. The additional vertical velocity v_y is proportional to this time, so $v_y = (a_y)t = \frac{qv_xB}{m}t$, where the vertical acceleration $a_y = \frac{F_y}{m}$ according to the Newton's second law of motion.

Putting these relationships together, we can calculate the ratio of velocity components:

$$\frac{v_y}{v_x} = \left(\frac{qv_xB}{m}\frac{L}{v_x}\right)\frac{1}{v_x} = \frac{q}{m}\frac{B^2L}{E}.$$

Suppose that the electric force provided the vertical acceleration instead of the magnetic force. The ratio would still be given by:

$$\frac{v_y}{v_x} = \left(\frac{qE}{m}\frac{L}{v_x}\right)\frac{1}{v_x} = \frac{q}{m}\frac{B^2L}{E}.$$

Note that both E and B, as well as L, are measurable quantities. The ratio $\frac{v_y}{v_x}$ is equal to the tangent of the angle at which the deflected electrons emerge out of the velocity selector region. Hence, Thomson's historical experiment gives us the charge-to-mass ratio of an electron:

$$\frac{q}{m} = \frac{E}{B^2L}\tan\theta.$$

## CHAPTER 7

Special Relativity

Albert Einstein's theory of special relativity came out of his deep thinking about electromagnetism, and we've been eager to tell you about this theory. It's about time.*

7.1 ABSOLUTENESS OF PHYSICAL LAWS

Consider two observers in two different inertial frames of reference, which means that they are moving at a constant velocity with respect to each other. For example, you, as one observer, may be riding on a train or an airplane that maintains a constant cruising speed without speeding up or slowing down. I, as the other observer, am just standing on the ground. You may have experienced a moment of a confusing sensation where you on a smoothly moving vehicle feel at rest, while other objects outside of the vehicle seem to be moving. In other words, I see you and your vehicle moving with velocity $\vec{v}$, while it would seem to you that I am moving with $-\vec{v}$.

A key idea of special relativity is that there is no absolute frame of reference. Hence, all inertial reference frames are equivalent, and the fundamental laws of physics, whether they were observed by you or me in two different inertial frames, should be the same. In principle, if the motion of your vehicle was perfectly smooth, it is fundamentally impossible to determine whether you and your vehicle are in motion or I am in motion in the opposite direction.

---

*Ha ha...

If you throw a ball straight up inside a smoothly moving vehicle, you would see that it goes up and down according to Newton's second law of motion $\vec{F}_{net} = m\vec{a}$. When I watch the motion of that ball from the outside, I would observe that the ball follows a parabolic trajectory $\vec{r}(t)$ that is different from the up-and-down motion $\vec{r'}(t)$ observed by you. However, the same law of mechanics explains the parabolic motion.

Even before Einstein, this intuitive and sensible idea that the laws of physics must be universal was accepted widely. The trajectory of the ball as a function of time can be described classically by the following two expressions, depending on the observer:

$$\vec{r}(t) = (vt)\hat{x} + (v_{0y}t - \frac{1}{2}gt^2)\hat{y}$$

$$\vec{r'}(t) = (v_{0y}t - \frac{1}{2}gt^2)\hat{y},$$

where the vehicle is moving at a constant velocity $\vec{v} = v\hat{x}$ and the initial vertical velocity of the ball is v_{0y}. The vertical acceleration due to the force of gravity in both inertial frames of reference is $-g$. A seemingly reasonable assumption is that time, as observed by you and me, is the same ($t' = t$), so we use t for both expressions. This assumption, as we will see in the next section, will be dramatically challenged and disproven by Einstein.

The second time derivatives of the above expressions are equal to each other, or $\frac{d^2\vec{r}}{dt^2} = \frac{d^2\vec{r'}}{dt^2} = -g\hat{y}$. The observed force of gravity, $-mg\hat{y}$, is expected to be identical in both inertial reference frames. Hence, with the classical, pre-Einsteinian assumption of absolute time and space, Newton's second law of motion is demonstrated to be universal. That is, $\vec{F}_{net} = m\vec{a}$ according to an observer in one inertial reference frame, and $\vec{F'}_{net} = m'\vec{a'}$ according to another observer in a different inertial reference frame.

Although these classical assumptions and results seem natural and logical, Einstein and other physicists have noticed some discrepancies in the context of electrodynamics. Consider the following scenario. You throw a ball at a horizontal velocity u inside of a vehicle moving with a constant velocity v. I am standing on the ground and watching the motion of the ball thrown by you. According to the classical analysis, I would see the ball moving with the velocity of $u + v$. If $u = 0.5c$ (at half the speed of light) and $v = 0.9c$ (90 percent of the speed of light), would the ball seem to be moving at a speed of $1.4c$, faster than the speed of light?

In the final chapter of this book, we will see that the laws of electromagnetism imply that the electromagnetic wave, which is known to us as light, moves at a constant speed of c in a vacuum. This result holds in any inertial frames of reference because every inertial frame should follow the same laws of electrodynamics. As a famous thought experiment goes, if you are moving at the speed of light and hold a mirror in front of you, would you see your own reflection? Would the light that has reflected off of you and is moving toward the mirror be able to catch up to the mirror, which is also moving at the speed of light? As an observer who sees you moving at c, would I observe the speed of light of your reflection to be $2c$ (answer: no) or still c (answer: yes)?

Einstein started with two fundamental postulates of special relativity: (1) the laws of physics are the same in all inertial frames of reference, and (2) the speed of light c is a constant, independent of the relative motion of the source. He showed that they naturally lead to quite remarkable conclusions about time and space.

7.2   TIME DILATION AND LENGTH CONTRACTION

Let's devise another thought experiment in which we measure the time it takes for light to travel a certain distance. Rather than holding the mirror in front of us, the mirror is attached to the ceiling of a moving train (a car, an airplane, or a rocket, as long as it moves at a constant velocity). A source and a detector of the light beam are installed on the floor, and when the light source is turned on, the light will travel to the ceiling, be reflected from the mirror, and return to the floor after a finite time. The train moves with a constant horizontal speed of v, so we will again consider two reference frames, S and S'. The reference frame S is for an observer on the ground, who sees the light beam moving vertically as well as horizontally along with the movement of the train. The reference frame S' is for an observer inside the train, who sees the light beam moving strictly vertically up and down. Each observer uses their own stopwatch to measure the time interval between the activation of the light source and its arrival back on the floor in their respective reference frames.

```
# Code Block 7.1

import numpy as np
import matplotlib.pyplot as plt
from matplotlib.patches import Rectangle
import sympy

# Time measurements in two inertial reference frames.

h = 1
vt = 1
ct = np.sqrt(h**2 + vt**2)
lw = 8

fig = plt.figure(figsize=(3,2))
plt.subplot(1,2,1)
plt.plot([0,vt],[0,0],color='gray')
plt.plot([vt,vt],[0,h],color='gray')
plt.quiver(0,0,vt,h,linewidth=lw,
            angles='xy',scale_units='xy',scale=1,color='k')
plt.title("(a) Frame S")
plt.text(vt+0.1,h/2,r"$h$")
plt.text(vt/2,-0.2,r"$v\Delta t/2$")
plt.text(vt/2-0.5,h/2,r"$c\Delta t/2$")
plt.xlim((-0.2,0.2+vt))
plt.ylim((0,h+0.1))
plt.xticks(())
plt.yticks(())
plt.axis('off')

plt.subplot(1,2,2)
plt.quiver(0,0,0,h,linewidth=8,
            angles='xy',scale_units='xy',scale=1,color='k')
plt.title("(b) Frame S'")
plt.text(-0.17,h/2,r"$c'\Delta t'/2$")
plt.text(+0.03,h/2,r"$h'$")
plt.xlim((-0.2,0.2))
plt.ylim((0,h+0.1))
plt.xticks(())
plt.yticks(())
plt.axis('off')
plt.tight_layout()
plt.savefig('fig_ch7_time_dilation.pdf')
plt.show()
```

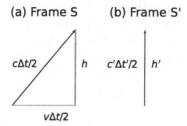

Figure 7.1

In S' where the source, the detector, and the mirror appear to be stationary, the relationship among h' (vertical distance between the floor and the ceiling), c' (speed of light), and $\Delta t'$ (time for the light beam to start from the source, hit the mirror on the ceiling, and return to the floor) is given by $2h' = c'\Delta t'$ or

$$\Delta t' = \frac{2h'}{c'}.$$

In S, the source, the detector, and the mirror are moving with speed v with respect to the observer on the ground. Hence, the trajectory of the light would take a diagonal path. The relationship among h (vertical distance), c (speed of light), v, and Δt (time for light to traverse the path) is given by the Pythagorean theorem, $(c\Delta t/2)^2 = (v\Delta t/2)^2 + (h)^2$. We can solve for Δt and obtain the following result.

$$\Delta t = \frac{2h}{\sqrt{c^2 - v^2}} = \frac{2h}{c\sqrt{1 - \beta^2}} = \gamma\frac{2h}{c},$$

where $\beta \equiv \frac{v}{c}$ (a unitless factor less than 1, since $c > v$) and $\gamma \equiv \frac{1}{\sqrt{1-\beta^2}}$ (a unitless factor larger than 1).

According to one of the postulates of special relativity, $c' = c$, or the speed of light is the same in all reference frames. Also, $h' = h$, because the measurement of the vertical distance, orthogonal to the relative velocity, would be the same in all reference frames. To justify this, you might imagine two paint brushes are attached to the outside of the moving train (frame S') such that they are spaced h' apart vertically. As the train moves with a constant horizontal speed v, each brush leaves a mark on a vertical pole fixed on the ground (frame S). The distance h between the paint marks on the pole would be equal to h'.

We can relate Δt and $\Delta t'$. Because $\frac{h'}{c'} = \frac{h}{c}$, $\Delta t = \gamma \Delta t'$. Because $\Delta t'$ is a time interval between two events happening at the same position, it is also known as proper time and often denoted as Δt_0. This implies that the time between two events measured in two different inertial reference frames is relative:

$$\Delta t = \gamma \Delta t_0.$$

This result is known as time dilation. Two observers in two different inertial reference frames are watching the same events: the light starts off from its source and arrives at a detector. The time interval between these events, as measured by an observer on the ground, would be longer than the time interval measured by an observer inside the train. In other words, the observer on the ground would conclude that a time-keeping device in the moving train must be running slowly. From the absoluteness of physical laws and the absoluteness of c, we arrived at the relativeness of time measurements.

Let us do a different thought experiment of measuring length in two distinct inertial reference frames. Imagine that we install a light source and a detector on one end of an object. On the other end, we place a mirror. The length of the object can be determined based on the time that the light takes to travel from the source to the mirror and back to the detector. This object is brought onto a moving train, and its length is measured by two observers: one inside the train (frame S') and the other on the ground (frame S).

Inside the moving train, the observer would see the object at rest, and this frame of reference S' is referred to as the rest frame of this object, and the length L_0 at its rest frame is called a proper length. We can relate the proper length L_0 and the travel time $\Delta t'$ of the light, as follows. The light traverses the full length of the object twice during its round trip, and each one-way trip takes the same amount of time in the rest frame, so $2L_0 = c\frac{\Delta t'}{2} + c\frac{\Delta t'}{2}$, where $\frac{\Delta t'}{2}$ is the one-way travel time. The speed of light is always c, according to the postulate of relativity. Hence,

$$\Delta t' = \frac{2L_0}{c}.$$

Measuring the length of a moving object is trickier. As the light travels from its source to the mirror during its one-way trip, it would have to travel an extra distance to catch up to the mirror that is also moving at v. On its return trip after being reflected from the mirror, the light would travel a shorter distance, because the detector would have moved toward the mirror. Let's denote the first one-way travel time to the mirror as Δt_1 and the return travel time from the mirror to the detector as Δt_2. Also, we will denote the length of the object according to the outside observer in the reference frame S as L.

During the first leg of the trip, according to an observer in S, the light would travel the distance of $L + v\Delta t_1$ over the time duration of Δt_1, so $\Delta t_1 = \frac{L+v\Delta t_1}{c}$. During the second leg of the trip, the light would travel the distance of $L - v\Delta t_2$ over Δt_2, so $\Delta t_2 = \frac{L-v\Delta t_2}{c}$. We can solve for Δt_1 and Δt_2 separately. Then, the total time for the round trip would be

$$\Delta t = \Delta t_1 + \Delta t_2 = \frac{L}{c-v} + \frac{L}{c+v} = \frac{2L}{c}\frac{1}{1-\frac{v^2}{c^2}}.$$

Now remember Δt and $\Delta t'$ in two reference frames are related by the factor γ, according to time-dilation, or $\Delta t = \gamma \Delta t'$. Hence, we can conclude that

$$L = \frac{L_0}{\gamma}.$$

The moving object looks shorter than its rest length, and this result is called length contraction.

```
# Code Block 7.2

# Length measurements in two inertial reference frames.

v = 0.6 # Try varying this value.
c = 1
y0 = 0.4
h = 0.3
gamma = 1/np.sqrt(1-v**2/c**2)
L0 = 1
L = L0/gamma

t1 = L/(c-v)
t2 = L/(c+v)

fig = plt.figure(figsize=(3,2))
plt.subplot(1,2,1)
ax = plt.gca()
ax.add_patch(Rectangle((0,y0),L,h,lw=1,alpha=0.3,
                edgecolor='k',facecolor='gray'))
ax.add_patch(Rectangle((v*t1,y0),L,h,lw=1,alpha=0.3,
                edgecolor='k',facecolor='gray',
```

```
                            linestyle='--'))
ax.add_patch(Rectangle((v*(t1+t2),y0),L,h,lw=1,alpha=0.3,
                        edgecolor='k',facecolor='gray'))
plt.quiver(0,0.5,L+v*t1,0,linewidth=8,
           angles='xy',scale_units='xy',scale=1,color='k')
plt.quiver(L+v*t1,0.6,-(L-v*t2),0,linewidth=8,
           angles='xy',scale_units='xy',scale=1,color='k')
plt.title("(a) Frame S")
plt.text(v*(t1+t2+0.1),0.62,r"$c\Delta t_2$")
plt.text(v*(t1+t2+0.1),0.52,r"$c\Delta t_1$")
plt.text(L/2,y0-0.1,r"$L$")

plt.ylim((0.2,0.8))
plt.xticks(())
plt.yticks(())
plt.axis('off')

plt.subplot(1,2,2)
ax = plt.gca()
ax.add_patch(Rectangle((0,y0),L0,h,lw=1,alpha=0.3,
                        edgecolor='k',facecolor='gray'))
plt.quiver(0,0.5,L0,0,linewidth=8,
           angles='xy',scale_units='xy',scale=1,color='k')
plt.quiver(L0,0.6,-L0,0,linewidth=8,
           angles='xy',scale_units='xy',scale=1,color='k')
plt.title("(b) Frame S'")
plt.text(L0/2,0.62,r"$c'\Delta t'/2$")
plt.text(L0/2,0.52,r"$c'\Delta t'/2$")
plt.text(L0/2,y0-0.1,r"$L_0$")

plt.ylim((0.2,0.8))
plt.xticks(())
plt.yticks(())
plt.axis('off')
plt.tight_layout()
plt.savefig('fig_ch7_length_contraction.pdf')
plt.show()
```

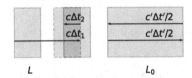

Figure 7.2

7.3 LORENTZ TRANSFORMATION

Different observers would specify the same event, a point in space and time, differently according to their own coordinate systems associated with their reference frames. The relationship between these specifications of spatial and temporal coordinates is called the coordinate transformation. If the coordinate systems are in different inertial reference frames, the coordinate transformation must be consistent with the results of time dilation and length contraction.

Suppose we have two Cartesian coordinate systems of inertial reference frames S and S', which are moving relative to each other at a constant velocity of v along the x-axis. According to an observer in S, the location and time of an event may be specified with the spatial coordinates (x, y, z) and the time coordinate t. The coordinates of the same event in S' would be (x', y', z') and t', according to an observer in S'. Let us assume that the origins of two coordinate systems, O and O', coincide at $t = t' = 0$. From the perspective of the observer in S at time t, the position of the origin of S' along the x-axis is at vt. Also, the horizontal distance between the origin O' and x' in S' would be length-contracted when observed in S. Therefore, the relationship between x', x, and t can be expressed as $x = vt + \frac{x'}{\gamma}$, or $x' = \gamma(x - vt)$.

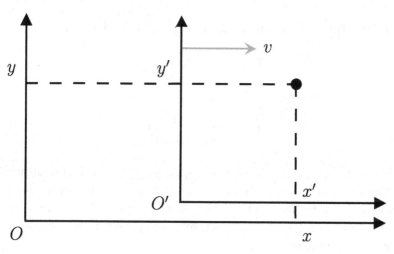

Figure 7.3

According to the observer in S', the coordinate system in S is moving with a constant velocity of $-v$. Therefore, we can obtain the inverse transformation, by writing x as x', t as t', x' as x, and v as $-v$. Hence, $x = \gamma(x' + vt')$. We can combine these two expressions to find the relationship between t', t, and x. The algebra is straightforward, as follows:

$$x = \gamma(x' + vt') = \gamma\left[\gamma(x - vt) + vt'\right] = \gamma^2(x - vt) + \gamma(vt'),$$

$$\gamma(vt') = \gamma^2(vt) + (1 - \gamma^2)x = \gamma^2 v\left[t + \left(\frac{1}{\gamma^2} - 1\right)\frac{x}{v}\right],$$

$$t' = \gamma\left[t + \left(\frac{1}{\gamma^2} - 1\right)\frac{x}{v}\right] = \gamma\left(t - \frac{v}{c^2}x\right).$$

The coordinates for the other two spatial dimensions are not affected by the horizontal movement (remember the previously mentioned thought experiment with two paint brushes?), so $y' = y$ and $z' = z$.

Hence, the relativistic coordinate transformation, known as Lorentz transformation, can be summarized as:

$$
\begin{aligned}
x' &= \gamma(x - vt) \\
y' &= y \\
z' &= z \\
t' &= \gamma\left(t - \frac{v}{c^2}x\right).
\end{aligned}
$$

The inverse transformation, going from S' to S, is given by:

$$
\begin{aligned}
x &= \gamma(x' + vt') \\
y &= y' \\
z &= z' \\
t &= \gamma\left(t' + \frac{v}{c^2}x'\right),
\end{aligned}
$$

which can be easily obtained by swapping the primed and unprimed variables and changing the sign of v. The Lorentz transformation formula explicitly illustrates the deep relationship and symmetry between space and time, and in relativity, we often refer to spacetime as a single, four-dimensional vector quantity.

From the Lorentz transformation, we can also derive the relativistic velocity addition formula. Consider an object moving with speed u' in the reference frame S'. The differential displacement of this object will be

dx' in dt', so $u' = \frac{dx'}{dt'}$. From the inverse transformation formula, we have $dx = \gamma(dx' + vdt')$ and $dt = \gamma(dt' + \frac{v}{c^2}dx')$. These differential displacements in space and time allow us to calculate the speed of an object according to an observer in S.

$$u = \frac{dx}{dt} = \frac{\gamma(dx' + vdt')}{\gamma(dt' + \frac{v}{c^2}dx')} = \frac{\frac{dx'}{dt'} + v}{1 + \frac{v}{c^2}\frac{dx'}{dt'}} = \frac{u' + v}{1 + \frac{vu'}{c^2}}.$$

This formula implies that if $u' = c$, $u = c$. In other words, the speed of light is constant regardless of the reference frame, which is consistent with the postulate of special relativity.

We can also easily deduce an inverse velocity addition formula, allowing us to calculate the speed of an object according to an observer in S', if u and v are known:

$$u' = \frac{u - v}{1 - \frac{vu}{c^2}}.$$

7.4 RELATIVISTIC MOMENTUM AND ENERGY

Relativistic momentum is defined as $\vec{p} = m\frac{d\vec{r}}{dt_0}$, where t_0 is the proper time measured in the rest frame of mass m. Defining the momentum as $m\frac{d\vec{r}}{dt}$ as in Newtonian mechanics is a bad choice since the length contraction and time dilation would make the old Newtonian definition of momentum a non-conserved quantity.

The time dilation result showed that $\Delta t = \gamma \Delta t_0$. Hence, the definition of relativistic momentum suggests that $\vec{p} = m\frac{d\vec{r}}{dt_0} = \gamma m\frac{d\vec{r}}{dt}$, where γ involves the speed of the object since the proper time t_0 is measured in the rest frame of the object itself. Energy is defined similarly as $E = mc^2\frac{dt}{dt_0} = \gamma mc^2$.

The relativistic definition of momentum $\vec{p} = m\frac{d\vec{r}}{dt_0}$ converts between different inertial reference frames according to the Lorentz transformation formula presented earlier, because dt_0 is the same in all reference frames and only the numerator requires transformation. However, for the Newtonian definition of momentum $m\frac{d\vec{r}}{dt}$, both the numerator and the denominator must be converted via Lorentz transformations between frames. This assures that if both relativistic momentum and energy are conserved in one inertial reference frame, they are conserved in another inertial reference frame.

For example, suppose $p = m\frac{dx}{dt_0}$ and $E = mc^2\frac{dt}{dt_0}$ are conserved in S. Then, in S', $p' = m\frac{dx'}{dt_0} = m\gamma(\frac{dx}{dt_0} - v\frac{dt}{dt_0})$ and $E' = mc^2\frac{dt'}{dt_0} = mc^2\gamma(\frac{dt}{dt_0} - \frac{v}{c^2}\frac{dx}{dt_0})$, where γ here involves the constant relative speed v between S and S', not the velocity of an object. The momentum p' and energy E' will also be conserved in S', because p and E are.

Let us make a few additional observations about these relativistic definitions of momentum and energy by considering a particle moving with velocity $\vec{u} = \frac{d\vec{r}}{dt} = (\frac{dx}{dt})\hat{x} + (\frac{dy}{dt})\hat{y} + (\frac{dz}{dt})\hat{z}$. Note $u^2 = (\frac{dx}{dt})^2 + (\frac{dy}{dt})^2 + (\frac{dz}{dt})^2$. First, when $u^2/c^2 \ll 1$ (i.e., non-relativistic regime), we can use the Taylor expansion for γ:

$$E = \gamma mc^2 = \frac{1}{\sqrt{1 - u^2/c^2}} mc^2 \approx \left(1 + \frac{u^2}{2c^2} + \cdots\right) mc^2 = mc^2 + \frac{1}{2}mu^2 \cdots,$$

where the leading term beyond a constant of mc^2 is the familiar (non-relativistic) kinetic energy formula, $\frac{1}{2}mu^2$.

Second, using the Lorenz transformation, we can show the following equality that is frame-independent:

$$
\begin{aligned}
E^2 - (\vec{p}c)^2 &= (\gamma mc^2)^2 - \gamma^2 m^2 c^2 \left[\left(\frac{dx}{dt}\right)^2 + \left(\frac{dy}{dt}\right)^2 + \left(\frac{dz}{dt}\right)^2\right] \\
&= (\gamma mc^2)^2 \left(1 - \frac{1}{c^2}u^2\right) \\
&= (\gamma mc^2)^2 \frac{1}{\gamma^2} \\
&= (mc^2)^2.
\end{aligned}
$$

A special case is a particle at rest with $\vec{p} = 0$, and the above equality reduces to $E = mc^2$, which is, of course, the world-famous physics equation.

Keeping in mind the ideas of relativistic momentum and energy, let us analyze a perfectly inelastic collision of two masses in one dimension, observed from two different reference frames. In a fixed reference frame S, the first mass m_1 moves at velocity u_1, while the second mass m_2 moves at velocity u_2 toward m_1. After the collision, two masses stick and move together. According to classical mechanics, the conservation of Newtonian momentum implies that

$$m_1 u_1 + m_2 u_2 = (m_1 + m_2)u_{\text{Newton}}$$

where u_{Newton} is the post-collision velocity of the combined mass (u_tot_newt in the code), so we have $u_{\text{Newton}} = (m_1 u_1 + m_2 u_2)/(m_1 + m_2)$.

However, unlike the momentum, some kinetic energy will be lost during such an inelastic collision event.

As a concrete example, if $m_2 = 2m_1$ and $u_1 = -u_2 = u$, $u_{\text{Newton}} = -\frac{1}{3}u$. Let us observe this collision event from another reference frame S' that is moving relative to frame S at a constant velocity of v. According to the relativistic velocity addition formula, m_1 and m_2 move at $\frac{u-v}{1-vu/c^2}$ and $\frac{-u-v}{1+vu/c^2}$ before collision respectively, and after collision, their velocity will be $\frac{-\frac{1}{3}u-v}{1+vu/3c^2}$. We expect the conservation of momentum, a universal law in physics, should remain valid in S'. However, as demonstrated by the following code block, the sum of Newtonian momenta before and after the collision in the new reference frame S' is not conserved.

According to relativity, both momentum and energy are conserved quantities, and their relations before and after the collision can be described as the following:

$$\gamma_1 m_1 u_1 + \gamma_2 m_2 u_2 = \gamma_t m_t u_t \text{ (Momentum conservation)}$$

$$\gamma_1 m_1 c^2 + \gamma_2 m_2 c^2 = \gamma_t m_t c^2 \text{ (Energy conservation)}$$

where $\gamma_1 = 1/\sqrt{1-u_1^2/c^2}$, $\gamma_2 = 1/\sqrt{1-u_2^2/c^2}$, and $\gamma_t = 1/\sqrt{1-u_t^2/c^2}$, since these relativistic factors involve the speed of each object. The above equations can be solved simultaneously for the post-collision velocity u_t and mass m_t. It is somewhat surprising that we have to solve for mass, given that we have always considered it an ever-constant value in classical physics. In contrast to the Newtonian case, the relativistic mass is not invariant, so we cannot assume $m_t = m_1 + m_2$. Our earlier discussion of $E = mc^2$ hints that the missing kinetic energy during the collision process manifests as the change in the mass.

The code block performs this analysis using the sympy module to solve for u_t and m_t (u_tot_eins and m_tot_eins in the code). The Einsteinian definition of momentum is conserved in both inertial frames of reference. If interested, you may insert a print() command to check the value of m_tot_eins after the collision.

```
# Code Block 7.3

# Perfectly inelastic collision between two masses

# The velocities (u, v) are in unit of c, so c = 1.
m1, m2 = 1, 2
u1, u2 = 0.8, -0.8

def add_velocity (u,v):
```

```
        return (u-v)/(1-u*v)

def momentum_eins (m,u):
    gamma = (1-u**2)**(-0.5)
    return gamma*m*u

def momentum_newt (m,u):
    return m*u

hw, hl = 0.03, 0.1 # size of arrowheads
s1, s2 = 50, 200 # size of the markers
c1, c2 = 'black', 'gray'
frame = ("(a) Frame S","(b) Frame S'")

fig,ax = plt.subplots(2,2,figsize=(3,2))
for i in range(2):

    if i==0: # Frame S with v=0
        # Newtonian case
        m_tot_newt = m1+m2
        u_tot_newt = (m1*u1 + m2*u2) / m_tot_newt

        # Einsteinian case
        # Solve for u_t and m_t using the conservations of
        # relativistic momentum and energy.
        m_t, u_t = sympy.symbols('m_t u_t')
        gamma_t = (1-u_t**2)**(-0.5)
        g1m1 = (1-u1**2)**(-0.5)*m1
        g2m2 = (1-u2**2)**(-0.5)*m2
        equations = [g1m1*u1+g2m2*u2-gamma_t*m_t*u_t,
                     g1m1+g2m2-gamma_t*m_t]
        solutions = sympy.solve(equations, m_t, u_t)
        m_tot_eins = float(solutions[0][0])
        u_tot_eins = float(solutions[0][1])

    else: # Frame S'
        v = 0.5
        # Velocities in a different reference frame
        u1 = add_velocity(u1,v)
        u2 = add_velocity(u2,v)
        u_tot_newt = add_velocity(u_tot_newt,v)
        u_tot_eins = add_velocity(u_tot_eins,v)

    p_beg_newt = momentum_newt(m1,u1) + momentum_newt(m2,u2)
    p_beg_eins = momentum_eins(m1,u1) + momentum_eins(m2,u2)
    p_end_newt = momentum_newt(m_tot_newt,u_tot_newt)
    p_end_eins = momentum_eins(m_tot_eins,u_tot_eins)
    p_newt = (p_beg_newt,p_end_newt)
    p_eins = (p_beg_eins,p_end_eins)
```

```
        print(frame[i])
        txt = 'momenta (before, after) = '
        print('  Newtonian %s(%3.2f, %3.2f)'%(txt,*p_newt))
        print('Einsteinian %s(%3.2f, %3.2f)'%(txt,*p_eins))
        print('')

        # Before collision
        ax[0,i].scatter(-1,0,s=s1,color=c1)
        ax[0,i].scatter(+1,0,s=s2,color=c2)
        ax[0,i].arrow(-1,0,u1,0,color=c1,
                      head_width=hw,head_length=hl)
        ax[0,i].arrow(+1,0,u2,0,color=c2,
                      head_width=hw,head_length=hl)

        # After collision
        ax[1,i].scatter(0,0,s=s2,color=c2)
        ax[1,i].scatter(0,0,s=s1,color=c1)
        ax[1,i].arrow(0,0,u_tot_eins,0,color=c1,
                      head_width=hw,head_length=hl)

        for j in range(2):
            ax[j,i].set_xlim((-1.2,1.2))
            ax[j,i].set_ylim((-0.2,0.2))
            ax[j,i].axis('off')

        ax[0,i].set_title(frame[i])

plt.tight_layout()
plt.savefig('fig_ch7_inelastic_collision.pdf')
plt.show()
```

(a) Frame S
  Newtonian momenta (before, after) = (-0.80, -0.80)
Einsteinian momenta (before, after) = (-1.33, -1.33)

(b) Frame S'
  Newtonian momenta (before, after) = (-1.36, -2.03)
Einsteinian momenta (before, after) = (-4.43, -4.43)

(a) Frame S   (b) Frame S'

Figure 7.4

7.5 RELATIVITY OF ELECTROMAGNETIC FORCES

Now, we are ready to discuss a seemingly paradoxical situation that beautifully illustrates the deep connection between the electric and magnetic forces. To understand how electricity and magnetism are intertwined through relative motion, let us consider two inertial reference frames. In the first reference frame S, there is a current-carrying wire and a positive charge q moving parallel to the straight wire at a distance d. The charge, moving in the same direction as the current at speed v, would experience the magnetic force of magnitude qvB, where B is the strength of the magnetic field due to current I in the wire. In the chapter on magnetism, we have already discussed that a straight wire with current I generates a magnetic field of magnitude $\frac{\mu_0 I}{2\pi d}$ at a distance d.

To make our analysis simpler, we may assume that this current is generated by the movements of positive and negative electrical charges in the wire moving in opposite directions at speed u. Hence, the linear charge density λ in the wire would have to satisfy $I = 2\lambda u$. The factor of 2 reflects the fact that positive and negative charges slide past each other in the wire, and each charge type contributes to half of the total current. In this reference frame, the positive charge density is equal to the negative charge density, so the wire itself is electrically neutral. The force exerted on q is purely magnetic, and its magnitude is given by:

$$F_B = qv\frac{\mu_0 \lambda u}{\pi d}.$$

The direction of force is given by the right-hand rule. If the current I and the charge velocity v are parallel to each other, the magnetic force points toward the wire.

In the second reference frame S', the charge q is at rest, so there would be no magnetic force on q. Hence, there seems to be a paradox. How can the charge q experience force in S, but none in S'? How could the presence of force be dependent on the frame of reference? As we will see below, the relativistic effect of length contraction resolves this paradox. In the rest frame of q, the positive charges, which are running alongside q in our arrangement, move slower than the negative charges, which are running in the opposite direction. As an extreme case, suppose both positive and negative charges move in the opposite direction at a speed identical to the speed of q. Then, according to an observer in S', the positive charges and q are at rest, while the negative charges are moving.

Due to length contraction, the spacing between the adjacent negative charges would be shorter than its proper length and the spacing between the positive charges. In other words, the charge density of the negative charges increases due to length contraction in S', so q will be exposed to a greater number of negative charges than positive charges and the wire is no longer electrically neutral unlike in S. There would be a net electric force on q. In this reference frame, the charge experiences a purely electric force.

These two different reference frames are illustrated in the following code block. By changing the speed u of the positive and negative charges or speed v_q of the charge q next to the wire, you can visualize the charge densities in the wire as observed in the reference frames of S and S'.

```
# Code Block 7.4

u = 0.6 # speed of the pos and neg charges (currents) in the wire.
v_q = 0.4 # speed of charge q
c = 1 # light speed (set to 1, so other speeds are relative to c).

plt.figure(figsize=(3,2))
step = 0.25 # spacing between charge symbols.
marg = 0.08 # spacing between the arrow and charge symbols.

for i in range(2):
    if i==0:
        v = 0
        title_str = "(a) Frame S"
    else:
        v = v_q # S' is the rest frame of q.
        title_str = "(b) Frame S'"

    # Relativistic velocity addition for
    # the pos and neg charges in the wire.
    u_pos = (+u-v)/(1-u*v/c**2)
    u_neg = (-u-v)/(1+u*v/c**2)

    # Length contraction
    gamma_pos = 1/np.sqrt(1-u_pos**2/c**2)
    gamma_neg = 1/np.sqrt(1-u_neg**2/c**2)
    range_pos = np.arange(-0.5,1+step/gamma_pos,step/gamma_pos)
    range_neg = np.arange(-0.5,1+step/gamma_neg,step/gamma_neg)

    plt.subplot(1,2,i+1)
    plt.title(title_str)
    plt.scatter(range_pos,np.zeros(len(range_pos))+marg,
            marker='+',color='k')
```

```
    plt.scatter(range_neg,np.zeros(len(range_neg))-marg,
              marker='_',color='k')
    plt.quiver(0,+3*marg,u_pos,0,scale=2,color='k')
    plt.quiver(0,-3*marg,u_neg,0,scale=2,color='k')

    # Plot the single charge q.
    plt.scatter(0,-0.5,marker='o',s=100,color='gray')
    plt.quiver(0,-0.5,v_q-v,0,scale=2,color='gray')
    plt.text(0,-0.8,'q')

    plt.xlim((-1,1))
    plt.ylim((-1.25,0.5))
    plt.axis('off')

plt.tight_layout()
plt.savefig('fig_ch7_relativistic_EM_force.pdf')
plt.show()
```

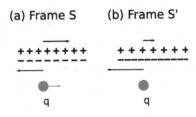

Figure 7.5

Let's do a more exact and careful analysis of this situation. We will denote the spacing between the adjacent charges in the wire in their rest frame, or the proper length, as L_0. According to an observer in S who sees the charges already in motion at speed u, this spacing between charges would be length-contracted and equal to $\frac{L_0}{\gamma_0}$, where $\gamma_0 = \frac{1}{\sqrt{1-u^2/c^2}}$. This length contraction of L_0 involves u, not v, because it is about the movement of the charges in the wire, not the movement of q. Correspondingly, if the charge density in the rest frames of the charges in the wire is denoted as λ_0, the charge density in S, where the charges are in motion, is $\lambda = \gamma_0 \lambda_0$.

What are the speeds of the moving charges in the rest frame of q? To answer this question, we have to use the velocity addition formula and write u' in terms of v and u:

$$u' = \frac{u - v}{1 - vu/c^2}.$$

The speed of the positive charges in the wire, as observed in S', is $u'_+ = \frac{u-v}{1-vu/c^2}$, and the speed of the negative charges is $u'_- = \frac{-u-v}{1+vu/c^2}$. This difference in speed leads to the different amount of length contraction for the positive and negative charges in S'. The length contraction factors for the positive and negative charges are: $\gamma_+ = \frac{1}{\sqrt{1-(u'_+/c)^2}}$ and $\gamma_- = \frac{1}{\sqrt{1-(u'_-/c)^2}}$. The respective charge densities are $\lambda_+ = +(\gamma_+\lambda_0)$ and $\lambda_- = -(\gamma_-\lambda_0)$, for the positive and negative charges, respectively. Due to the unequal densities, the wire would be electrically charged with a non-zero total charge density of $\lambda_T = \lambda_+ + \lambda_-$, when observed in S'.

The electric field strength at a distance d away from an infinite wire with charge density λ_T is given by $E = \frac{|\lambda_T|}{2\pi\epsilon_0 d}$, which can be quickly verified with Gauss's law. Hence, the magnitude of the electric force exerted on the stationary charge q in S' is:

$$F'_E = \frac{q|\lambda_T|}{2\pi\epsilon_0 d} = \frac{q\lambda_0|\gamma_+ - \gamma_-|}{2\pi\epsilon_0 d} = \frac{q\lambda|\gamma_+ - \gamma_-|}{\gamma_0 2\pi\epsilon_0 d}.$$

The expression $|\gamma_+ - \gamma_-|$ appearing in F'_E can be simplified further. To get an idea of how it may be simplified, we use the sympy module, which suggests that the two terms may be combined with a common denominator of $\sqrt{(c+u)(c-u)(c+v)(c-v)}$.

```
# Code Block 7.5

u, v, c = sympy.symbols('u v c')
u_pos = (+u-v)/ (1-u*v/c**2)
u_neg = (-u-v)/ (1+u*v/c**2)
gamma_pos = 1/sympy.sqrt(1-u_pos**2/c**2)
gamma_neg = 1/sympy.sqrt(1-u_neg**2/c**2)

display(sympy.combsimp(gamma_pos))
display(sympy.combsimp(gamma_neg))
```

$$\frac{1}{\sqrt{\frac{(-c+u)(-c+v)(c+u)(c+v)}{c^4-2c^2uv+u^2v^2}}}$$

$$\frac{1}{\sqrt{\frac{(-c+u)(-c+v)(c+u)(c+v)}{c^4+2c^2uv+u^2v^2}}}$$

Therefore,

$$
\begin{aligned}
|\gamma_+ - \gamma_-| &= \left| \frac{(c^2 - uv) - (c^2 + uv)}{\sqrt{(c+u)(c-u)(c+v)(c-v)}} \right| \\[2mm]
&= \left| \frac{-2uv}{\sqrt{(c+u)(c-u)(c+v)(c-v)}} \right| \\[2mm]
&= \left| -\frac{2uv}{c^2} \frac{1}{\sqrt{1 - u^2/c^2}} \frac{1}{\sqrt{1 - v^2/c^2}} \right| \\[2mm]
&= \frac{2uv}{c^2} \gamma_0 \gamma,
\end{aligned}
$$

where $\gamma_0 = \frac{1}{\sqrt{1-u^2/c^2}}$ and $\gamma = \frac{1}{\sqrt{1-v^2/c^2}}$. The electric force expressed in terms of u and v becomes

$$
F'_E = \gamma \frac{q\lambda uv}{c^2 \pi \epsilon_0 d}.
$$

What would be the direction of the force? This can be inferred from examining the polarity of the wire observed in S'. Due to the relatively fast-moving negative charges, we have $\gamma_- > \gamma_+$, which in turn leads to $\lambda_T < 0$ (negatively charged wire). Hence, the positive charge q will experience an attractive electric force toward the wire, the same direction as the magnetic force F_B in the frame S.

Finally, in order to compare two forces in two different reference frames, namely F'_E in S' and F_B in S, we need to do one more conversion from S' to S. That is, if the upward force in S' is identified as F'_E, what would that be in S, whose relative velocity with respect to q is $-v$?

An observer in one particular reference frame can describe a force as a time rate of the relativistic momentum according to Newton's second law of motion. The observer will see this change in the momentum with respect to his or her own time in that frame. Therefore, the force is defined as a derivative of the momentum with respect to t, not proper time t_0, as $\vec{F} = \frac{d\vec{p}}{dt}$.

With this definition of force in relativistic dynamics, we may return to our original question about a moving charge near a wire. The above determination of the force pointing upward prompts us to look at the momentum in the same upward direction, perpendicular to the horizontal path of q in S and the relative motion of the frame S'. Let's denote this vertical direction as y and y'-axis, while the horizontal direction as x and x'-axis, in S and S', respectively. In S', $F'_E = \frac{dp'_y}{dt'}$ and $p'_y = m\frac{dy'}{dt_0}$. In S, $F_E = \frac{dp_y}{dt}$ and $p_y = m\frac{dy}{dt_0}$. Note that the spatial dimension perpendicular to the relative motion of the frame is invariant under the coordinate transformation, that is, $y = y'$. Hence, $p_y = m\frac{dy}{dt_0} = m\frac{dy'}{dt_0} = p'_y$.

Let's find the relationship between the respective vertical forces responsible for the momentum changes in each frame: $F_y = \frac{dp_y}{dt}$ and $F'_y = \frac{dp'_y}{dt'}$. Since, according to the Lorentz transformation, the differential time is given by $dt = \gamma(dt' + \frac{v}{c^2}dx')$,

$$
\begin{aligned}
F_y &= \frac{dp_y}{dt} \\
&= \frac{dp'_y}{\gamma(dt' + \frac{v}{c^2}dx')} \\
&= \frac{\frac{dp'_y}{dt'}}{\gamma\left(1 + \frac{v}{c^2}\frac{dx'}{dt'}\right)} \\
&= \frac{1}{\gamma}\frac{dp'_y}{dt'} \\
&= \frac{1}{\gamma}F'_y
\end{aligned}
$$

where we used $\frac{dx'}{dt'} = 0$ in the rest frame of q.

In other words, we can find the expression for F_E, by multiplying $\frac{1}{\gamma}$ to F'_E:

$$
F_E = \frac{1}{\gamma}F'_E = \frac{q\lambda uv}{c^2\pi\epsilon_0 d} = qv\frac{\mu_0\lambda u}{\pi d} = F_B.
$$

Here, we used the fact that the speed of light is given by $c \equiv \frac{1}{\sqrt{\epsilon_0\mu_0}}$, which will be discussed in the later chapter on the electromagnetic wave. The magnetic force observed in S is identical to the electric force observed in S'.

This completes our analysis of the relativity of electromagnetic forces, revealing that electricity and magnetism are relativistic manifestations of each other. When we hold an ordinary piece of a magnet, we are witnessing a relativistic phenomenon without needing to approach the speed of light.

## CHAPTER 8

Potential

8.1 ELECTRIC POTENTIAL

Consider going along an electric field through a particular path. Along this path, let's add up the strength of the field pointing in the same direction. Mathematically, this summation can be expressed as $\int_{\text{Path}} \vec{\mathbf{E}} \cdot d\vec{\mathbf{l}}$, where $d\vec{\mathbf{l}}$ denotes the infinitesimal segment along the path. This is similar to the calculation of flux, which is the integration of the strength of the electric field parallel to the normal of some specified surface.

Electric potential V is defined as

$$V = -\int_{\text{Path}} \vec{\mathbf{E}} \cdot d\vec{\mathbf{l}}.$$

One volt (V) of potential is equivalent to one joule per coulomb. Unfortunately, the symbol V is associated with several different quantities and may cause some confusion. V may refer to electric potential, the unit volt of potential, volume, or velocity in lower case. We sometimes have to distinguish them based on context. The negative sign means that we are going against the direction of the electric field. That is, the voltage definition goes "up" or against the direction of the field, rather than "down."

What is the meaning and significance of V? As a motivating example, consider a well-studied example from classical mechanics: an object with mass m under the influence of gravity near the surface of the earth. When I lift this object over a vertical distance h by doing mechanical work, it

DOI: 10.1201/9781003397496-8

has moved in the opposite direction (there is the negative sign again) of the gravitational force and gained gravitational potential energy, mgh, because

$$
\begin{aligned}
\text{Gravitational Potential Energy} \ &= \ -\text{Work} \\
&= \ -\int \vec{F} \cdot d\vec{l} \\
&= \ -\int_0^h (-mg)dy = mgh.
\end{aligned}
$$

Because the electric force $\vec{F} = q\vec{E}$, the definition of electric potential leads to

$$
\begin{aligned}
V \ &= \ -\int \vec{E} \cdot d\vec{l} \\
&= \ -\frac{1}{q}\int \vec{F} \cdot d\vec{l} \\
&= \ \text{Electrical Potential Energy divided by } q.
\end{aligned}
$$

Conversely, we can define the electric field in terms of the electric potential. The differential of the potential dV along an infinitesimal segment $d\vec{l} = dx\hat{x} + dy\hat{y} + dz\hat{z}$ is $dV = \frac{\partial V}{\partial x}dx + \frac{\partial V}{\partial y}dy + \frac{\partial V}{\partial z}dz$, which can be expressed as $\nabla V \cdot d\vec{l}$, or the gradient of the potential function, so the electric potential becomes

$$
V = \int_{\text{Path}} dV = \int_{\text{Path}} \nabla V \cdot d\vec{l}.
$$

Our original definition of the potential was $V = -\int_{\text{Path}} \vec{E} \cdot d\vec{l}$, so we identify that $\vec{E} = -\nabla V$. In other words, the three-dimensional slope of the electric potential V is the negative of the electric field $\vec{E}$.

The concept of electric potential is analogous to the gravitational potential energy divided by m, or gh in the above example. Just as gh indicates how much gravitational potential energy a unit mass would possess when it is lifted by height h, the electric potential V indicates how much electrical potential energy a unit charge would possess when it is moved under the presence of the electric field between the initial and final positions.

In the case of the gravitational potential energy, the earth's surface often serves as a convenient point of reference where the gravitational potential energy is set to zero. Similarly, we often choose to treat a point at infinity as our reference point for the electric potential, so that $V(\infty) = 0$. With this reference point of zero potential at infinity, we can treat the electric potential at one point in space as being equal to the amount of work to bring a unit electrical charge from infinity to that particular position under the presence of the electric field. Sometimes, a reference point other than infinity can be more convenient. For example, when we are dealing with electrical circuits, we often define "ground" which serves as an infinite source or sink of electrical charges and assign zero potential to it. A biological cell also has electrical potential difference across its membrane, and it is customary to consider the outside of the cell to be at zero potential, making the intracellular region at negative potential.

The gravitational force is conservative. The amount of mechanical work done on an object as it is moved within the gravitational field is independent of its path of movement. The change in the gravitational potential energy of an object is only determined by the change of its position, as long as no non-conservative force like air drag or friction is involved. In other words, we only need h, the change in height, to calculate the change in the gravitational potential energy. It does not matter whether the mass m was raised straight up or in a zig-zag path. Similarly, the electric potential is path-independent. You can bring a charge along very different paths, but the change in the electric potential of a charge is determined solely by two positions: the charge's starting and ending positions.

A ball on a hilly landscape would roll down the hill, trading its potential energy with kinetic energy. A charge placed at a higher electric potential also rolls down the gradient of the electric potential landscape. Using the analogy of pressure can also be helpful to describe potential. As the difference in water pressure causes the flow of water, the difference in electric potential causes the flow of electric charges. An electric charge can be either positive or negative, unlike the mass of an object which is always a positive quantity. A positive charge rolls "down" the electric potential landscape, while a negative charge rolls "up" and moves in the opposite direction of the positive charge.

8.2 ELECTRIC CIRCUIT

In Thomson's experiment, two oppositely charged plates created an electric field, pointing from the positive to the negative plate, and there was an electrical potential difference between them. A positive electric charge placed between the plates would accelerate along the electric field between the electric potential difference. A negative electric charge would accelerate in the opposite direction. An electric circuit works by the same principle. A typical battery has the capacity to provide the electrical potential difference due to the differential affinity of electrons of two different materials at each end of the electrodes. Such a capacity is called emf^* or $\mathcal{E}$. When various electronic components, such as a resistor or a capacitor, are connected via conducting wires to a battery, the battery's emf offers the necessary "push" for the charges to flow through the circuit.

A resistor controls the amount of electric current in a circuit. When two ends of a resistor are placed under different potentials, the charges will flow, and the rate of the charge flow I depends on the potential difference ΔV across the resistor and the resistor's resistance R. The resistance depends on various factors, such as the material, temperature, shape, and size (cross-sectional area and length) of a resistor. The resistance is measured with a unit of ohm (Ω). A resistor, during a typical operation of a circuit, has a constant resistance and exhibits a linear relationship such that $\Delta V = IR$, which is called Ohm's law.

Another common circuit element is a capacitor, which stores and releases charges, acting as a temporary reservoir, and it can affect the temporal behavior of a circuit. A pair of conducting plates that are separated by a non-conducting medium is a prototypical configuration of a capacitor. The capacitance C is the measure of the capacitor's capacity to hold charges, and it is determined mainly by the size of the plates and the gap between them. Capacitance can be boosted by increasing the size of the plates and by filling the gap with a dielectric material. The capacitance is measured with a unit of farad (F).

When emf is applied across two capacitor plates, electrical charges start to flow. Free electrons can move and accumulate on one plate, charging it negatively, while the other plate becomes positively charged due to the

---

*It used to be called an electromotive force but is a historical misnomer, because emf is not really a force.

corresponding loss of free electrons. This separation of charges comes at the expense of doing work, because moving an electron to a negatively charged plate requires overcoming the repulsive force between the electron and the plate. It becomes progressively more difficult to add additional charges. As the two plates accumulate equal-and-opposite electrical charges, the capacitor gains a potential difference ΔV, which is proportional to the amount of charge accumulation q. The capacitance is equal to the amount of charge that can be held by a capacitor per unit potential difference, or 1 farad = 1 coulomb per volt, so $\Delta V = q/C$.

8.3 RC CIRCUIT

Let us consider a simple circuit where a resistor with resistance R and a capacitor with capacitance C are connected in series to a battery with *emf* of $\mathcal{E}$. In the previous section, we discussed how each component acts individually, and here we analyze the behavior of the circuit with two conservation laws: the conservation of electric charge and the conservation of energy, which are also known as Kirchhoff's first and second rules in the context of electrical circuits. The first rule states that the current flowing into any point in a circuit always equals the current flowing out of that point. This is true even when a point includes multiple branches. In other words, an electrical charge does not magically appear nor disappear within the circuit, or the electrical charge is conserved. The second rule states that the sum of the potential changes around any loop within a circuit is always zero. In other words, when an electrical charge travels through the circuit and returns to its starting point, the net change of its electrical potential energy is zero. It means that, while a charge is traversing a loop, any gain in the electrical potential energy is offset by the loss, due to the work performed by the charge or energy contributed to different parts of the circuit. The energy does not magically appear or disappear in the circuit, or the energy is conserved.

In the context of an RC circuit, the first rule implies that the number of electrical charges that move to or from the capacitor plate will be equal to the number passing through the resistor at any moment in time, so we can use the same symbol q to refer to the charges at R and C. The second rule implies that the battery's *emf* is equal to the sum of the potential drops at R and C. The work done by the battery is split into the work of moving electrical charges against the electrical resistance and the work of charging up the capacitor.

$$\mathcal{E} = IR + \frac{q}{C} = R\frac{dq}{dt} + \frac{q}{C}$$

with $I = \frac{dq}{dt}$. This differential equation with respect to time has a solution of $q(t) = C\mathcal{E}(1 - e^{-t/RC})$, which shows the charging behavior of a capacitor with a characteristic time constant of RC. The time-derivative of $q(t)$ is $\frac{dq}{dt} = \frac{\mathcal{E}}{R}e^{-t/RC}$, which shows the exponentially decreasing current in the circuit, as the capacitor charges up.

This differential equation can be analyzed with the `sympy` module, too. In the following code block, the ordinary differential equation `potential_eqn` contains a symbol q, which is defined as a function. The solution is obtained by the `sym.dsolve()` function, with the initial condition of $q(0) = 0$ (i.e., the capacitor is uncharged at the beginning). Once $q(t)$ is found, the current in the circuit is determined by taking its time-derivative with `sym.diff()`.

```
# Code Block 8.1

import numpy as np
import matplotlib.pyplot as plt
import sympy as sym

# Analyze RC circuit, symbolically.

# Define emf, resistance, capacitance, and time.
E, R, C, t = sym.symbols('E R C t')

# Define charge as a function.
q = sym.symbols('q', cls=sym.Function)

# Define the ordinary differential equation.
potential_eqn = sym.Eq(R*q(t).diff(t)+q(t)/C,E)

# Solve the ordinary differential equation.
q_func = sym.dsolve(potential_eqn,ics={q(0): 0},simplify=True).rhs

# Solve for current.
I_func = sym.diff(q_func,t)

# Display the results.
print('charges q(t) =', q_func)
print('current I(t) =', I_func)
```

```
charges q(t) = C*E - C*E*exp(-t/(C*R))
current I(t) = E*exp(-t/(C*R))/R
```

The following code block illustrates this capacitor-charging process. The negative charges, or electrons, accumulate on the capacitor plate connected to the battery's negative terminal, and the other plate develops the positive polarity. This charge accumulation process slows down exponentially, indicated by the decrease in current flow through the resistor. The voltage changes in the resistor and the capacitor, IR and $\frac{q}{C}$, respectively, add up to the battery's emf.

```
# Code Block 8.2

# Visualizing an RC circuit with cartoon sketches.

def draw_RC_circuit(i,t_now,V):

    # Plotting parameters.
    lw = 1 # linewidth.
    fs = 8 # fontsize.
    V_R, V_C = V

    # Positions of circuit elements:
    # battery, capacitor, resistor, wires.
    bat_x = [(-1.5,-0.8),(1.5,0.8)]
    bat_y = [(2,1.5),(2,1.5)]
    cap_x = [(8,8),(12,12)]
    cap_y = [(2,1),(2,1)]
    res_x = [2.5,7.5,7.5,2.5]
    res_y = [4.5,4.5,5.5,5.5]
    wire_x = [(0,0,0,7.5,10,10,0),(0,0,2.5,10,10,10,10)]
    wire_y = [(2,1.5,5,5,5,1,-2),(5,-2,5,5,2,-2,-2)]

    plt.plot(bat_x, bat_y, color='gray', linewidth=lw)
    plt.plot(cap_x, cap_y, color='gray', linewidth=lw)
    plt.fill(res_x, res_y, edgecolor='gray', facecolor='none',
            linewidth=lw)
    plt.plot(wire_x, wire_y, color='gray', linewidth=lw)
    plt.text(2,1,r'$\mathcal{E}$',fontsize=fs)
    plt.text(6,1,r'$C$',fontsize=fs)
    plt.text(4,3,r'$R$',fontsize=fs)

    # Number and positions of charges on each capacitor plate.
    # For a visualization purpose (it's just a cartoon),
    # increase the number of charge markers linearly.
```

```
    q_num = 2*i
    x = np.linspace(8,12,q_num)
    y_p = 2.5*np.ones(len(x)) # location of positive markers
    y_m = 0.5*np.ones(len(x)) # location of negative markers
    plt.scatter(x,y_p,s=10,marker='+',color='black',linewidths=lw)
    plt.scatter(x,y_m,s=10,marker='_',color='black',linewidths=lw)

    plt.yticks([])
    plt.xticks([])
    plt.axis('equal')
    plt.text(14,4,r"at t = %2.1f $RC$"%(t_now),fontsize=fs)
    plt.text(14,2,r"$\Delta V$ at $R$ = %3.2f $\mathcal{E}$"%(V_R),
            fontsize=fs)
    plt.text(14,0,r"$\Delta V$ at $C$ = %3.2f $\mathcal{E}$"%(V_C),
            fontsize=fs)
    plt.box(on=False)
    return

# Simulate RC circuit at different times.
R_value, C_value = 1, 2
RC = R_value*C_value
t_range = np.array([0,0.5,1,1.5])*RC

fig = plt.figure(figsize=(4,5))
for i, t_now in enumerate(t_range):
    # Calculate charge and current at t_now.
    value_dict = {E:1,R:R_value,C:C_value,t:t_now}
    q_now = q_func.subs(value_dict).evalf()
    I_now = I_func.subs(value_dict).evalf()
    V_R, V_C = I_now*R_value, q_now/C_value
    plt.subplot(len(t_range),1,i+1)
    draw_RC_circuit(i,t_now/RC,(V_R,V_C))

plt.tight_layout()
plt.savefig('fig_ch8_RC_circuit_cartoon.pdf')
plt.show()
```

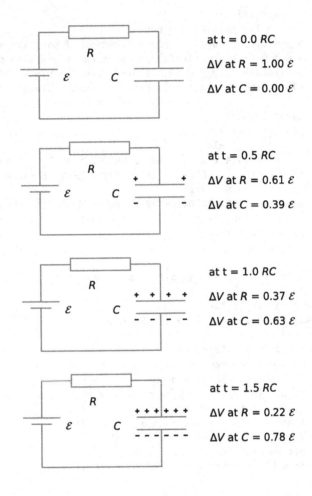

Figure 8.1

The following code block calculates the behavior of an RC circuit with a $200\ k\Omega$ resistor, a $2\ \mu F$ capacitor, and a battery with an *emf* of $12\ V$.

```
# Code Block 8.3

# Calculate voltage changes in R and C over time.
fig = plt.figure(figsize=(3,3))
t_range = np.arange(0,20.1,0.2)
q = np.zeros(len(t_range))
I = np.zeros(len(t_range))

E_value = 12 # 12 volt battery
```

```
R_value = 200*10**3 # 200 kOhm resistor
C_value = 20*10**(-6) # 20 micro farad capacitor
for i, t_value in enumerate(t_range):
    value_dict = {E:E_value,R:R_value,C:C_value,t:t_value}
    q[i] = q_func.subs(value_dict).evalf()
    I[i] = I_func.subs(value_dict).evalf()

plt.plot(t_range,q/C_value,color='black',linestyle='solid')
plt.plot(t_range,I*R_value,color='black',linestyle='dotted')
plt.legend((r'$\Delta V$ (capacitor)',
            r'$\Delta V$ (resistor)'),
           framealpha=1)
plt.xlabel('t (sec)')
plt.ylabel('Volts')
plt.yticks(np.array([0,0.5,1])*E_value)
plt.xticks(np.array([0,0.5,1])*np.max(t_range))
plt.tight_layout()
plt.savefig('fig_ch8_RC_circuit.pdf')
plt.show()
```

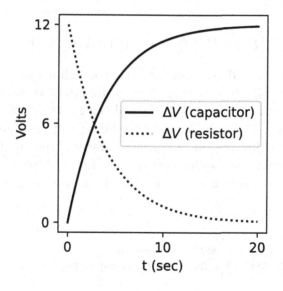

Figure 8.2

8.4 RC CIRCUIT UNDER AC

The behavior of an RC circuit becomes more interesting when we have a time-varying power source. For example, what happens when the *emf* is sinusoidal with frequency ω and amplitude A?

$$\mathcal{E}(t) = A \sin(\omega t) = R\frac{dq(t)}{dt} + \frac{q(t)}{C}.$$

We can tweak the previous code blocks to examine an interesting behavior of an RC circuit when an alternating current (AC) is applied. The following plots demonstrate that there is a frequency-dependent phase difference between the driving emf, and q and I. The voltage across each circuit element is also frequency-dependent. The amplitude of $\Delta V(t)$ $(= q(t)/C)$ at the capacitor is small when ω is large. In other words, the high-frequency voltage is attenuated, and the low-frequency voltage signal passes through the circuit without much reduction in its amplitude. Hence, this circuit configuration is also known as a low-pass filter. High-pass and band-pass filters are possible with other types of circuit elements.

A particular application of a low-pass filter is illustrated in the fourth plot. In this example,

$$\mathcal{E}(t) = (1 - f)A \sin(\omega t) + fA \sin(4\omega t),$$

where the fraction f determines how much of high-frequency noise is embedded in the $\mathcal{E}(t)$. In this example, high-frequency is chosen to be four times the base frequency ω. When the driving emf is a sum of multiple frequencies, an RC circuit filters out the high-frequency components and outputs a smoothed-out, low-frequency version of the input. The values of R and C determine the cut-off frequency. In acoustics, a low-pass filter may be used to emphasize the bass sounds, as a subwoofer does.

```
# Code Block 8.4

# Analyze an AC RC circuit.
R_value = 200*10**3 # 200 kOhm resistor
C_value = 20*10**(-6) # 20 micro farad capacitor
A_value = 2

# Different cases
which_case = 1
if which_case==1:
    w_value = 0.5
    noise_fraction = 0
if which_case==2:
    w_value = 2
    noise_fraction = 0
```

```
if which_case==3:
    w_value = 1
    noise_fraction = 0
if which_case==4:
    w_value = 1
    noise_fraction = 0.3

# Define resistance, capacitance, amplitude, frequency, and time.
R, C, A, w, t = sym.symbols('R C A w t')

q = sym.symbols('q', cls=sym.Function)
E = sym.symbols('E', cls=sym.Function)
E = (1-noise_fraction)*A*sym.sin(w*t)
E = E + noise_fraction*A*sym.sin(4*w*t)

# Define the ordinary differential equation.
potential_eqn = sym.Eq(R*q(t).diff(t)+q(t)/C,E)

# Solve the ordinary differential equation.
q_func = sym.dsolve(potential_eqn,ics={q(0): 0},simplify=True).rhs
I_func = sym.diff(q_func,t)

# Calculate voltage changes in R and C over a time interval.
fig = plt.figure(figsize=(3,3))
t_range = np.arange(0,40.1,0.2)
q = np.zeros(len(t_range))
I = np.zeros(len(t_range))

for i, t_value in enumerate(t_range):
    value_dict = {A:A_value,w:w_value,R:R_value,C:C_value,t:t_value}
    q[i] = q_func.subs(value_dict).evalf()
    I[i] = I_func.subs(value_dict).evalf()

emf = I*R_value + q/C_value
plt.plot(t_range,emf,color='#CCCCCC',linestyle='solid')
plt.plot(t_range,q/C_value,color='#000000',linestyle='solid')
plt.plot(t_range,I*R_value,color='#AAAAAA',linestyle='dotted')
plt.legend((r'$\mathcal{E}$ (emf)',
            r'$\Delta V$ (capacitor)',
            r'$\Delta V$ (resistor)'),
           loc='lower left',framealpha=1)
plt.xlabel('t (sec)')
plt.ylabel('Volts')
title_str = r'$\omega$ = %2.1f, $f$ = %2.1f'%(w_value,noise_fraction)
plt.title(title_str)
plt.yticks(np.array([-1,0,1])*A_value)
plt.ylim(np.array([-1.1,1.1])*A_value)
plt.xticks(np.array([0,0.5,1])*np.max(t_range))
plt.tight_layout()
```

```
plt.savefig('fig_ch8_RC_circuit_AC_case%d.pdf'%which_case)
plt.show()
```

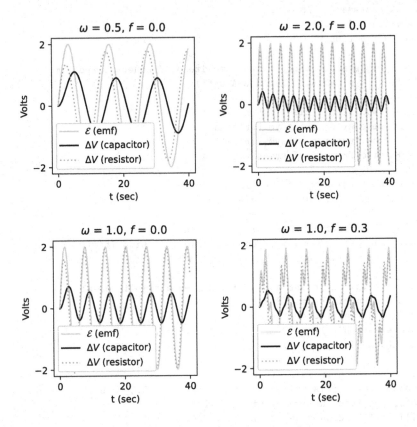

Figure 8.3: RC circuit with different driving frequency ω and noise level.

The physics of an RC circuit underlies how our brain works, too. Neurons are surrounded by and filled with a liquid solution with various ions (most notably, Na^+, K^+, Ca^{++}, and Cl^-). An active ion-pumping process creates differential ionic concentrations and electrical potential difference across the cellular membrane, which is usually kept at approximately 70 mV. The membrane, therefore, acts as a capacitor that separates the interior and exterior of the cell. There are many gated pores, or ion channels, in the membrane, too. There are voltage-gated, ligand-gated, and stretch-gated channels, each of which can be further specialized for different ion species. The ion channels open and close, letting and stopping the flow of the ions across the cellular membrane, so they act as variable resistors. When these ion channels act in a co-ordinated manner, the ionic currents produce a rapid and significant fluctuation in the cell's electrical potential, called an action potential or spike. That is one of the ways how the neurons transmit signals to each other. Another communication method is via chemical diffusion.

8.5 MAGNETIC POTENTIAL

It is rather intuitive to imagine the electric potential as a scalar function with specific magnitudes at each position in space, similar to the gravitational potential. The steepest slope in the electric potential landscape matches the direction of the electric field, and the electric force on a charge points along that direction. The magnetic force, on the other hand, pushes a charge perpendicular to both the velocity and the magnetic field. The magnetic force and the magnetic field do not point in the same direction, unlike the electric force and the electric field. Therefore, we have to work with a different form of a potential function in the case of the magnetic field.

An electric charge, which is a scalar quantity without direction, sets up the electric potential V, which is itself a scalar function. Similarly, an electric current, a vector quantity with both magnitude and direction, is the source of the magnetic field and sets up the magnetic potential, which is a vector function. The magnetic potential is often denoted with $\vec{A}$. The relationship between the magnetic field and the magnetic potential is:

$$\vec{B} = \nabla \times \vec{A}.$$

The electric potential is related to the potential energy of a charge, and the potential difference tells us the work done by the electric field per unit charge. However, the magnetic potential does not carry such physical interpretation because the magnetic field does not perform any work. The magnetic vector potential takes on important theoretical role when both electric and magnetic fields are considered in light of quantum mechanics, which is beyond the scope of this book. For now, we may consider the magnetic potential $\vec{A}$ as another way of obtaining the magnetic field $\vec{B}$, just as the electric field $\vec{E}$ can be described as the gradient of electric potential V. Sometimes, the magnetic potential allows us to calculate the magnetic field more easily than the Biot-Savart law.

8.6 MATH: SCALAR AND VECTOR POTENTIALS

In this section, we will introduce a beautiful mathematical theory behind the relationship between fields and potentials. The fundamental theorem of vector calculus, also known as Helmholtz theorem, states that a vector field like $\vec{E}$ or $\vec{B}$ can always be represented as, or decomposed into, a combination of gradient and rotational fields, such as ∇V and $\nabla \times \vec{A}$, respectively. The gradient field is always curl-free, and the rotational field is always divergence-free.

Theorems about decomposition or representation have foundational roles in mathematics and physics. For example, the fundamental theorem of arithmetic states that any integer can be represented as a product of prime numbers. The Fourier theorem states that a periodic function can be decomposed into a series of sinusoidal functions. With the knowledge that a mathematical object (in our case, a vector field) can be represented with other mathematical objects with unique properties (like curl-free or divergence-free), we can more easily analyze, synthesize, understand, and manipulate their behaviors.

Much of our scientific and cognitive efforts of understanding the universe consist of breaking up a complex system or phenomena into more-approachable components. For example, in chemistry, the structure of a molecule is analyzed based on the properties of its constituent atoms. In biology, a cell is studied based on the complex interactions among its constituent parts such as nucleic acids, proteins, and organelles. In the standard model of particle physics, a theory of fundamental forces is considered based on a small set of elementary particles: quarks, leptons, elementary bosons, and their antiparticles. In economics, the behavior of a market is analyzed based on the different incentives and actions of its constituent players, such as consumers and producers. Now, let's get into more detail about the scalar and vector potentials for electromagnetic fields.

The relationship between the field and the potential, $\vec{E} = -\nabla V$ and $\vec{B} = \nabla \times \vec{A}$, comes from more general mathematical statements about a vector field. A gradient vector field $\vec{K}$ with a zero curl can always be expressed as a gradient of some scalar function V, where $\vec{K} = \nabla V$.

With $\vec{K} = \nabla V$, the curl of $\vec{K}$ is:

$$
\begin{aligned}
\nabla \times \vec{K} &= \nabla \times (\nabla V) \\
&= \left(\frac{\partial^2 V}{\partial y \partial z} - \frac{\partial^2 V}{\partial z \partial y} \right) \hat{x} + \left(\frac{\partial^2 V}{\partial z \partial x} - \frac{\partial^2 V}{\partial x \partial z} \right) \hat{y} + \left(\frac{\partial^2 V}{\partial x \partial y} - \frac{\partial^2 V}{\partial y \partial x} \right) \hat{z}.
\end{aligned}
$$

Because the variables x, y, and z are independent, the order of differentiation does not affect these mixed second partial derivatives, so $\nabla \times \vec{K} = 0$, as expected for a gradient field.

A rotational vector field $\vec{K}$ with a zero divergence can always be expressed as a curl of some vector function $\vec{A}$, where $\vec{K} = \nabla \times \vec{A}$. With $\vec{A} = A_x \hat{x} + A_y \hat{y} + A_z \hat{z}$, the divergence of $\vec{K}$ is:

$$\nabla \cdot \vec{K} = \nabla \cdot \left(\nabla \times \vec{A} \right)$$
$$= \left(\frac{\partial^2 A_z}{\partial x \partial y} - \frac{\partial^2 A_y}{\partial x \partial z} \right) + \left(\frac{\partial^2 A_x}{\partial y \partial z} - \frac{\partial^2 A_z}{\partial y \partial x} \right) + \left(\frac{\partial^2 A_y}{\partial z \partial x} - \frac{\partial^2 A_x}{\partial z \partial y} \right).$$

As before, because the successive differentiation is interchangeable, the sum of these terms is equal to zero.

Let's be more specific about the form of these vector fields. Is the electric field of a point charge curl-free? In other words, is it irrotational, and is the following statement true?

$$\nabla \times \frac{1}{4\pi\epsilon_0} \frac{Q}{r^2} \hat{r} = 0.$$

The answer is yes, as shown below.

```
# Code Block 8.5

# Demonstrate that curl of E equals zero.
# No need to consider constants likes: pi, Q, epsilon_0

x, y, z = sym.symbols('x y z')

# E-field from single charge
Ex = x/(x**2+y**2+z**2)**(3/2)
Ey = y/(x**2+y**2+z**2)**(3/2)
Ez = z/(x**2+y**2+z**2)**(3/2)

# Calculation of curl
curlx = (sym.diff(Ez,y)-sym.diff(Ey,z))
curly = (sym.diff(Ex,z)-sym.diff(Ez,x))
curlz = (sym.diff(Ey,x)-sym.diff(Ex,y))

print("Curl of E-field =",[curlx,curly,curlz])
```

Curl of E-field = [0, 0, 0]

If the point charge is not at the origin, we can consider a shifted coordinate system of $x' = x + C$, where C is a constant. The ∇' operation in this new coordinate system will involve $\frac{\partial}{\partial x'}$, but it is equal to $\frac{\partial}{\partial x}$, because the added constant C does not affect the derivative. Therefore, $\nabla \times \vec{E} = 0$, regardless of the position of the charge.

The curl-free nature of the electric field can also be understood via Stoke's theorem. We have already discussed that the line integral of the electric field is path-independent and only depends on the starting and ending positions. Therefore, the line integral of the electric field over a closed path must equal zero, and Stoke's theorem ensures that $\oint_{\text{Contour}} \vec{E} \cdot d\vec{s} = \iint_{\text{Area}} (\nabla \times \vec{E}) \cdot d\vec{a} = 0$. Since this must be true for any arbitrary surface area for the double integral, its integrand should always be zero, or $\nabla \times \vec{E} = 0$, and $\vec{E}$ can be expressed with a scalar function as $\vec{E} = -\nabla V$.

Because of the superposition principle, the total electric field of any charge distribution can be written as a sum of electric fields from many different point charges. Each of these electric fields will have a zero curl. Hence, the total electric field will also be curl-free. In other words, $\nabla \times \vec{E}_{\text{total}} = \nabla \times \vec{E}_1 + \nabla \times \vec{E}_2 + \cdots = 0 + 0 + \cdots = 0$.

Is the magnetic field from a steady current divergence-free or rotational? Let's consider the magnetic field from an infinitely long wire (assuming zero thickness) with the current, I flowing along $+\hat{z}$. We can determine this magnetic field either by considering Ampere's law or by using $\vec{B}_{\text{wire}}$ obtained for a wire with a finite length L and taking the limit of L going to infinity:

$$\vec{B} = \frac{\mu_0 I}{2\pi r}(-\sin\phi\hat{x} + \cos\phi\hat{y}),$$

where the polar angle ϕ is given as $\tan^{-1}(y/x)$. Is the divergence of this field zero? In other words, is the following statement true?

$$\nabla \cdot \frac{\mu_0 I}{2\pi r}(-\sin\phi\hat{x} + \cos\phi\hat{y}) = 0.$$

The answer is yes, as shown below.

```
# Code Block 8.6


# Demonstrate that divergence of B is zero.
# No need to consider constants likes: pi, I, mu_0


x, y, z = sym.symbols('x y z')
phi = sym.atan2(y, x)


# B-field from single charge
```

```
Bx = -sym.sin(phi)/(x**2+y**2)**(1/2)
By = +sym.cos(phi)/(x**2+y**2)**(1/2)
Bz = 0

# Calculation of divergence
div = sym.diff(Bx,x)+sym.diff(By,y)+sym.diff(Bz,z)

print("Divergence of B-field =",div)
```

Divergence of B-field = 0

The magnetic field from an infinitely long wire is divergence-free.

How about other shapes of current, such as a circular loop? The Biot-Savart law tells us that a magnetic field $\vec{B}(x, y, z)$ from a current source of an arbitrary shape with a current density $\vec{J}(x_s, y_s, z_s)$ can be given as a volume integral over the current source:

$$\vec{B}(x, y, z) = \frac{\mu_0}{4\pi} \iiint \frac{\vec{J}(x_s, y_s, z_s) \times \vec{r}}{r^3} dx_s dy_s dz_s$$

where the displacement vector between the source and observation points is $\vec{r} = (x - x_s)\hat{x} + (y - y_s)\hat{y} + (z - z_s)\hat{z}$ and $r = \sqrt{(x - x_s)^2 + (y - y_s)^2 + (z - z_s)^2}$.

The divergence of this integral form of the magnetic field is:

$$\nabla \cdot \vec{B}(x, y, z) = \frac{\mu_0}{4\pi} \iiint \nabla \cdot \left[\frac{\vec{J}(x_s, y_s, z_s) \times \vec{r}}{r^3} \right] dx_s dy_s dz_s.$$

Here, the ∇-operator, $\nabla = \frac{\partial}{\partial x}\hat{x} + \frac{\partial}{\partial y}\hat{y} + \frac{\partial}{\partial z}\hat{z}$, is placed inside the volume integral since the divergence is evaluated with respect to the observation point (x, y, z), not the source point (x_s, y_s, z_s). Also, the integrand can be re-expressed as

$$\nabla \cdot \left[\frac{\vec{J}(x_s, y_s, z_s) \times \vec{r}}{r^3} \right] = \frac{\vec{r}}{r^3} \cdot \left[\nabla \times \vec{J}(x_s, y_s, z_s) \right] - \vec{J}(x_s, y_s, z_s) \cdot \left(\nabla \times \frac{\vec{r}}{r^3} \right)$$

by applying the product rule for the divergence of the cross-product.

Because the current density is only a function of x_s, y_s and z_s, the first term, where the differentiation is taken with respect to x, y and z, is zero. We will evaluate $\nabla \times \frac{\vec{r}}{r^3}$ in the second term with the following code block.

```
# Code Block 8.7

# Curl of a displacement vector over distance cubed.

x, y, z, xs, ys, zs = sym.symbols('x y z xs ys zs')

# distance
r_cubed = (sym.sqrt((x-xs)**2+(y-ys)**2+(z-zs)**2))**3
rx = (x-xs)/r_cubed
ry = (y-ys)/r_cubed
rz = (z-zs)/r_cubed

# Calculation of curl
curlx = (sym.diff(rz,y)-sym.diff(ry,z)).simplify()
curly = (sym.diff(rx,z)-sym.diff(rz,x)).simplify()
curlz = (sym.diff(ry,x)-sym.diff(rx,y)).simplify()

print("Curl of displacement vector over distance cubed =")
print([curlx,curly,curlz])
```

```
Curl of displacement vector over distance cubed =
[0, 0, 0]
```

We showed that $\nabla \times \frac{\vec{r}}{r^3}$ always equals zero. This can be easily obtained by noticing $\frac{\vec{r}}{r^3} = -\nabla\left(\frac{1}{r}\right)$ and recalling that the curl of a gradient of a scalar function is zero. We may also apply the intuition that the curl represents the amount of circulation of a vector field and that the radial vector field does not have any circulation. Finally, with both terms in the integrand being zero, we conclude that $\nabla \cdot \vec{B} = 0$ always. Furthermore, there exist a vector potential $\vec{A}$, whose curl defines the magnetic field vector, or $\vec{B} = \nabla \times \vec{A}$.

One might ask whether there exists a unique scalar or vector function, V or $\vec{A}$, for a given gradient or rotational vector field $\vec{K}$, respectively. In our study of electromagnetism, this corresponds to expecting a unique electric potential for a given electric field and a unique vector potential for a given magnetic field. The answer is no. As an antiderivative $F(x)$ of a function $f(x)$ is only unique up to a constant (that is, a function $F(x) + C$ for any constant C can be an antiderivative of $f(x)$), the scalar and vector potential functions of $\vec{K}$ are not unique.

Given a particular scalar function V for a gradient field $\vec{K}$, we can add a constant C to form a different scalar function $V' = V + C$ that also satisfies $\nabla V' = \nabla(V + C) = \nabla V + \nabla C = \vec{K}$, because $\nabla C = 0$. For a rotational vector field $\vec{K}$ with a vector function $\vec{A}$, we can form a new vector function by adding a gradient of an arbitrary scalar function V, or $\vec{A}' = \vec{A} + \nabla V$. It will also satisfy $\nabla \times \vec{A}' = \nabla \times (\vec{A} + \nabla V) = \nabla \times \vec{A} + \nabla \times \nabla V = \vec{K}$, because $\nabla \times \nabla V = 0$.

We have thus discussed that a static electric field can always be represented in a curl-free way, with a gradient of a scalar function, and that a static magnetic field can always be represented in a divergence-free way, with a curl of a vector function. These representations are not unique, but non-uniqueness also means that we can tailor these functions and choose the most convenient representation for a situation at hand. This flexibility allows us to simplify the mathematical expressions of the electromagnetic fields and exploit the symmetry of a given problem without altering the underlying physics.

8.7 POISSON'S EQUATION

Thus far, we have been mostly dealing with the first derivatives of scalar and vector functions and just saw a few examples of the second derivatives, where ∇ was applied twice. Of particular importance is the divergence of gradient of a scalar function: $\nabla \cdot \nabla V = \frac{\partial^2 V}{\partial x^2} + \frac{\partial^2 V}{\partial y^2} + \frac{\partial^2 V}{\partial z^2}$, which is also written as $\nabla^2 V$. The second derivative operator ∇^2 is often called Laplacian.

Let's consider a class of differential equations with the Laplacian operator:

$$\nabla^2 V(x, y, z) = f(x, y, z),$$

which is known as Poisson's equation. A special version of Poisson's equation, where $f(x, y, z) = 0$, is called Laplace's equation.

If the scalar function V vanishes at infinity, or $V(\infty) \to 0$, a solution to Poisson's equation can be expressed as:

$$V(x, y, z) = -\frac{1}{4\pi} \iiint \frac{f(x', y', z')}{\sqrt{(x - x')^2 + (y - y')^2 + (z - z')^2}} dx' dy' dz'.$$

The proof of this result is beyond the scope of this book, as it involves other advanced mathematical topics, such as the Dirac delta function. In the remaining sections of this chapter, we will use this result to explore some examples of electromagnetic potentials.

8.8 EXAMPLE: ELECTRIC POTENTIAL OF CONTINUOUS CHARGE DISTRIBUTION

This second-order differential equation is important in electrostatics (and other branches of physics like thermodynamics), because $\nabla V = -\vec{E}$ and $\nabla \cdot \vec{E} = \frac{\rho}{\epsilon_0}$, so $\nabla \cdot \nabla V = \nabla^2 V = -\frac{\rho}{\epsilon_0}$. In other words, the solution to Poisson's equation is the electric potential for a given charge distribution ρ. In free space without any charge ($\rho = 0$), we have $\nabla^2 V(x, y, z) = 0$.

For charge density ρ, the corresponding electric potential V can be expressed as

$$V(x, y, z) = \frac{1}{4\pi\epsilon_0} \iiint \frac{\rho}{r} dx_s dy_s dz_s$$

where $r = \sqrt{(x - x_s)^2 + (y - y_s)^2 + (z - z_s)^2}$ is the distance between the point of observation (x, y, z) and the source point (x_s, y_s, z_s) of electric charge density. This integral will be evaluated over a volume where the electric charge density is non-zero.

The intuition for this solution is as follows. In the earlier chapters, we have established that the total flux of the electric field through an enclosed surface is directly proportional to the amount of electric charge enclosed within the volume (this is Gauss's law). Hence, the spread, or the divergence, of the electric field is determined by the electric charge density ρ. Another observation we have made earlier is that the electric field strength depends on the distance from the source charge, decreasing as $\frac{1}{r^2}$ in three-dimensional space. Since the electric field is related to the negative gradient, or the rate of spatial change, of the electric potential, the electric potential itself can be calculated by adding up electric charges while adjusting their relative contributions based on their positions. The charges located farther away contribute less to the electric potential by a factor of $\frac{1}{r}$.

As a simple example, imagine there is a point charge q located at the origin, or $(x_s, y_s, z_s) = (0, 0, 0)$. The volume integral that encloses the origin reduces to q, and the electric potential at (x, y, z) beyond the origin is given by:

$$V(x, y, z) = -\frac{1}{4\pi\epsilon_0} \frac{q}{\sqrt{x^2 + y^2 + z^2}} = -\frac{q}{4\pi\epsilon_0 r},$$

where $r = \sqrt{x^2 + y^2 + z^2}$. (At the origin where $r = 0$, we have to deal with infinite charge density, which can be handled with the formulation of the Dirac delta function. However, we can sidestep this discussion by only considering the region where the electric potential is finite.) The negative gradient of V, or $-\frac{dV}{dr}\hat{r}$, is indeed equal to the electric field $\vec{E} = \frac{q}{4\pi\epsilon_0 r^2}\hat{r}$, as expected.

If electric charges are distributed in space, we again rely on the super-position principle and approximate the charge distribution as the sum of point charges. For each point charge, its electric field is equal to the negative gradient of the electric potential. Hence, $\vec{E}_{\text{total}} = \vec{E}_1 + \vec{E}_2 + \cdots = -\nabla V_1 - \nabla V_2 - \cdots = -\nabla V_{\text{total}}$, and $V_{\text{total}} = V_1 + V_2 + \cdots$.

When the charge is distributed along a straight line, the electric potential can be calculated as a line integral. For example, consider a rod of length L, which is uniformly charged with Q and is lying along the x-axis. Its linear charge density is $\lambda = Q/L$. We can find the electric potential at distance z away from the center of the rod:

$$
\begin{aligned}
V(x = 0, y = 0, z) &= \frac{1}{4\pi\epsilon_0} \int_{\text{Rod}} \frac{\lambda}{r} dx_s \\
&= \frac{\lambda}{4\pi\epsilon_0} \int_{-L/2}^{+L/2} \frac{1}{\sqrt{z^2 + x_s^2}} dx_s.
\end{aligned}
$$

The following code block evaluates the above integral with the sympy module.

```
# Code Block 8.8

# Electric potential of a thin wire.

Q, e, L, x, z = sym.symbols('Q \epsilon_0 L x z')
charge_density = Q/L

# Integral form of Poisson equation solution.
d = sym.sqrt(z**2+x**2)
V = sym.integrate(charge_density/(4*sym.pi*e*d), (x,-L/2,L/2))

print('Electric potential V(z) =')
display(V.nsimplify())
```

Electric potential $V(z)$ =

$$\frac{Q\,\mathrm{asinh}\left(\frac{L}{2z}\right)}{2\pi L\epsilon_0}$$

The inverse hyperbolic sine function, asinh(), can be written differently as:

$$\mathrm{asinh}\left(\frac{L}{2z}\right) = \ln\left(\frac{L}{2z} + \sqrt{1+\left(\frac{L}{2z}\right)^2}\right),$$

and it is plotted below.

```python
# Code Block 8.9

# inverse hyperbolic sine function in a logarithmic form

x = np.arange(0,5,0.1)
y1 = np.arcsinh(x)
y2 = np.log(x + np.sqrt(1+x**2))

fig = plt.figure(figsize=(3,3))
plt.plot(x,y1,linewidth=5,color='#CCCCCC')
plt.plot(x,y2,linewidth=1,color='#000000')
plt.xticks((0,1,2,3,4,5))
plt.yticks((0,1,2))
plt.legend(('arcsinh(x)',r'$\ln(x + \sqrt{1+x^2})$'))
plt.axis('equal')
plt.tight_layout()
plt.savefig('fig_ch8_asinh.pdf')
plt.show()
```

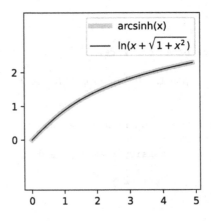

Figure 8.4

We can examine the limiting behaviors of the electric potential at the very near and far distances of the rod. When $z \gg L$ or $\frac{L}{2z} \ll 1$, asinh() behaves linearly, or $\text{asinh}(\frac{L}{2z}) \rightarrow \frac{L}{2z}$. Hence, the potential $V(z) \rightarrow \frac{Q}{4\pi\epsilon_0 z}$. This is identical to the potential from a single charge Q located at the origin, whose electric field at z is $\vec{E} = \frac{Q}{4\pi\epsilon_0 z^2}\hat{z}$. The electric potential at z due to this single charge is given as: $V = -\int_\infty^z \vec{E} \cdot d\vec{l} = \frac{Q}{4\pi\epsilon_0 z}$. In other words, this finite rod would look like a point charge at a large distance compared to its length.

At another limit of $z \ll L$ or $\frac{L}{2z} \gg 1$, this solution reduces to $V(z) = \frac{Q}{2\pi L\epsilon_0} \ln\left(\frac{L}{z}\right)$. We can take its negative gradient to determine the electric field with charge density λ:

$$\vec{E} = -\nabla V = -\frac{\partial}{\partial z}\left(\frac{\lambda}{2\pi\epsilon_0}\ln\left(\frac{L}{z}\right)\right)\hat{z} = \frac{\lambda}{2\pi\epsilon_0 z}\hat{z}.$$

This is identical to the electric field of an infinite, uniformly charged rod because, with the observation point very close to the wire, the wire would appear quite long. One can quickly verify this result with Gauss's law, using a cylindrical Gaussian surface. We also point out an interesting observation that the electric field of an infinite line in two dimensions, which was examined in an earlier chapter, is a constant value, whereas, in three dimensions, it depends on the distance, as shown here.

Here is another example. Consider a uniformly charged ring with a radius of R and total charge Q, placed at the origin on the xy-plane. The charge density of the ring is $\lambda = \frac{Q}{2\pi R}$. We can find the electric potential along the z-axis that goes through the ring's center. The solution of Poisson's equation can be evaluated as a line integral along the ring as

$$\begin{aligned} V(z) &= \frac{1}{4\pi\epsilon_0}\int_{\text{Ring}} \frac{\lambda}{\sqrt{z^2 + R^2}}dl \\ &= \frac{1}{4\pi\epsilon_0}\int_0^{2\pi} \frac{\lambda}{\sqrt{z^2 + R^2}}Rd\phi \\ &= \frac{1}{4\pi\epsilon_0}\frac{Q}{\sqrt{z^2 + R^2}}, \end{aligned}$$

where the differential length dl of the ring is $Rd\phi$ and the integration is performed over ϕ. When z is far away from the ring, the potential from the ring will be identical to that of the single charge Q located at the origin, with $V(z) = \frac{Q}{4\pi\epsilon_0 z}$.

As another example, consider a uniformly charged disc with radius R and total charge Q, placed at the origin on the xy-plane. The surface charge density of the disc is $\sigma = \frac{Q}{\pi R^2}$. The solution of Poisson's equation can be evaluated as a surface integral.

$$
\begin{aligned}
V(z) &= \frac{1}{4\pi\epsilon_0} \iint_{\text{Disc}} \frac{\sigma}{\sqrt{r^2 + z^2}} r dr d\phi \\
&= \frac{1}{4\pi\epsilon_0} \int_0^{2\pi} d\phi \int_0^R \frac{\sigma}{\sqrt{r^2 + z^2}} r dr \\
&= \frac{\sigma}{2\epsilon_0} \int_0^R \frac{r dr}{\sqrt{r^2 + z^2}} \\
&= \frac{\sigma}{2\epsilon_0} \sqrt{r^2 + z^2} \Big|_0^R \\
&= \frac{\sigma}{2\epsilon_0} (\sqrt{R^2 + z^2} - z).
\end{aligned}
$$

Let's take a limiting case where the disc extends to infinity, or $R \to \infty$. The above expression can be further simplified in terms of $\frac{z}{R}$.

$$
\begin{aligned}
V(z) &= \frac{\sigma R}{2\epsilon_0} \left(\sqrt{1 + \left(\frac{z}{R}\right)^2} - \frac{z}{R} \right) \\
&= \frac{\sigma R}{2\epsilon_0} \left(1 + \frac{1}{2}\left(\frac{z}{R}\right)^2 + \cdots - \frac{z}{R} \right) \\
&= \frac{\sigma}{2\epsilon_0} \left(R - z + \frac{z^2}{2R} + \cdots \right).
\end{aligned}
$$

The electric field of an infinite plane can be calculated by taking the negative derivative of the above expression with respect to z and then taking the limit of $R \to \infty$.

$$
\vec{E} = -\frac{dV}{dz}\hat{z} = \frac{\sigma}{2\epsilon_0}\hat{z}.
$$

which is independent of the distance from the plane and consistent with the result obtained in the earlier chapter.

As our last example of the electric potential, consider a sphere with radius R at the origin. It is uniformly charged with a charge density of $\rho = Q/(\frac{4}{3}\pi R^3)$. The electric potential from this sphere can be worked out better in the spherical coordinates. You can find the details on the spherical coordinate system in the appendix section of this book. Thanks to the spherical symmetry, we just need to consider the radial direction in this problem and will conveniently assume that the observation point is on the z-axis. The electric potential as a function of distance r from the center of the sphere will be given by:

$$V(r) = \frac{1}{4\pi\epsilon_0} \int_0^{2\pi} \int_0^{\pi} \int_0^R \frac{\rho}{\sqrt{r^2 + r_s^2 - 2rr_s\cos\theta_s}} r_s^2 \sin\theta_s dr_s d\theta_s d\phi_s,$$

where $\sqrt{r^2 + r_s^2 - 2rr_s\cos\theta_s}$ is the distance between the observation point at $(x, y, z) = (0, 0, r)$ and the source point. The coordinates of a source point is $(x_s, y_s, z_s) = (r_s\sin\theta_s\cos\phi_s, r_s\sin\theta_s\sin\phi_s, r_s\cos\theta_s)$. The code block below performs the symbolic calculation of this integral and plots the result for a unit sphere $(R = 1)$.

```
# Code Block 8.10

# Electric potential for spherical charge distribution.

R, r = sym.symbols('R r', positive=True)
rho = 1/((4/3)*sym.pi*R**3)
theta_s, phi_s, r_s = sym.symbols('\theta_s \phi_s r_s',
                                  positive=True)

# Integral form of Poisson equation solution.
d = sym.sqrt(r**2 + r_s**2 - 2*r*r_s*sym.cos(theta_s))
V = sym.integrate((r_s**2)*sym.sin(theta_s)/d,
              (theta_s,0,sym.pi),(phi_s,0,2*sym.pi),(r_s,0,R))
V = rho/(4*sym.pi)*V.simplify()
#display(V)

# Plot of electric potential inside and outside of the sphere.
r_range = np.arange(0.1,3.1,0.1)
norm_pot = np.zeros(len(r_range))

for k, r_value in enumerate(r_range):
    norm_pot[k] = 4*np.pi*V.subs({R:1,r:r_value}).evalf()

fig = plt.figure(figsize=(4,3))
plt.plot(r_range,norm_pot,label=r'$V_{sphere}$',
        color='black',linestyle='solid',linewidth=1)
plt.plot(r_range,1/r_range,label='1/r',
        color='gray',linestyle='dotted',linewidth=4)
plt.xlabel('r')
plt.ylabel(r'norm. potential ($\frac{Q}{4\pi \epsilon_0} = 1$)')
plt.legend()
plt.xticks(np.array([0,1,2,3]))
```

```
plt.yticks(np.array([0,1,2]))
plt.ylim(np.array([0,2.5]))
plt.tight_layout()
plt.savefig('fig_ch8_sphere.pdf')
plt.show()
```

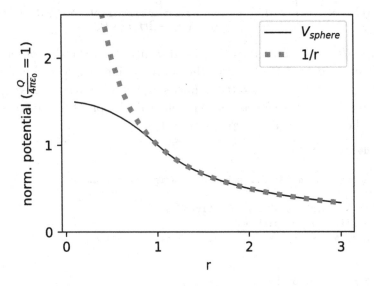

Figure 8.5

There are two distinct regions in this example: the inside $(r < R)$ and outside $(r > R)$ of the charged sphere. Outside of the sphere, $V(r > R) = \frac{Q}{4\pi\epsilon_0 r}$, which is the same as the electric potential of a single charge. Inside of the sphere, $V(r < R) = \frac{Q}{8\pi\epsilon_0 R}\left(3 - \left(\frac{r}{R}\right)^2\right)$.

8.9 EXAMPLE: MAGNETIC POTENTIAL OF CONTINUOUS CURRENT DISTRIBUTION

For the magnetic vector potential, we will use the following mathematical identity about the curl of the curl of a vector field: $\nabla \times (\nabla \times \vec{K}) = \nabla(\nabla \cdot \vec{K}) - \nabla^2\vec{K}$. Let's take the curl of the magnetic field:

$$\nabla \times \vec{B} = \nabla \times (\nabla \times \vec{A}) = \nabla(\nabla \cdot \vec{A}) - \nabla^2\vec{A} = \mu_0\vec{J},$$

where we have used $\vec{B} = \nabla \times \vec{A}$ and the differential form of the Ampere's law.

As discussed previously, the vector potential for the magnetic field is not uniquely determined, and we can choose a solution that may simplify the equation by adding or subtracting an arbitrary gradient of a scalar function. We will pick the divergence-free vector potential $\vec{A}$, such that $\nabla \cdot \vec{A} = 0$. Then, the above expression reduces to

$$\nabla^2 \vec{A} = -\mu_0 \vec{J},$$

which is a vector version of Poisson's equation. Its solution can be expressed as:

$$\vec{A}(x, y, z) = \frac{\mu_0}{4\pi} \iiint \frac{\vec{J}}{r} dx_s dy_s dz_s$$

where r is the distance between the point where the potential is evaluated and the current source point.

We can decompose this expression into three Poisson's equations in three-dimensional space with one Poisson's equation for each spatial dimension. In the cartesian coordinates, the three components of the magnetic potential $\vec{A}$ can be obtained by solving three equations: $\nabla^2 A_x = -\mu_0 J_x$, $\nabla^2 A_y = -\mu_0 J_y$, and $\nabla^2 A_z = -\mu_0 J_z$.

Let's consider a wire of length L with current I flowing in the direction of $+\hat{x}$. Assuming a thin wire, the current density is given by $\vec{J} dx_s dy_s dz_s = +I dx_s \hat{x}$. Because $\vec{J}$ has neither y nor z-component, $\vec{A}$ would have only x-component A_x. The solution to Poisson's equation at a distance z from the middle of the wire is

$$A_x(z) = \frac{\mu_0}{4\pi} \int_{-L/2}^{+L/2} \frac{I}{\sqrt{z^2 + x_s^2}} dx_s.$$

We had come across the same integral when we dealt with the electric potential of a charged wire. We can simply swap μ_0 with $\frac{1}{\epsilon_0}$ and I with λ. The magnetic vector potential of a current-carrying wire is:

$$\vec{A}(z) = \frac{\mu_0 I}{2\pi} \ln \left[\frac{L}{2z} + \sqrt{1 + \left(\frac{L}{2z}\right)^2} \right] \hat{x}.$$

When $z \ll L$, this solution becomes $\vec{A}(z) = \frac{\mu_0 I}{2\pi} \ln\left(\frac{L}{z}\right) \hat{x}$. At this limit, let's find the magnetic field by taking the curl of this vector potential.

$$\vec{B} = \nabla \times \vec{A}(z) = \frac{\partial A_x(z)}{\partial z} \hat{y} = -\frac{\mu_0 I}{2\pi z} \hat{y},$$

which is, as expected, identical to the result obtained with Ampere's law and the right-hand rule.

As another example of the magnetic vector potential, consider a current-carrying ring with radius R centered at the origin on the xy-plane. The current flow can be expressed as $I d\vec{l} = -IR d\phi \sin\phi \hat{x} + IR d\phi \cos\phi \hat{y}$. The magnetic potential only has the x and y components. At distance z away from the ring's center, A_x and A_y are given by:

$$A_x(z) = \frac{\mu_0}{4\pi} \int_0^{2\pi} \frac{-IR \sin\phi}{\sqrt{z^2 + R^2}} d\phi = 0$$

$$A_y(z) = \frac{\mu_0}{4\pi} \int_0^{2\pi} \frac{IR \cos\phi}{\sqrt{z^2 + R^2}} d\phi = 0.$$

The above integrals look complicated, but they are just integrals of a sinusoidal function over its full period, so they are equal to zero. Is it surprising that the magnetic vector potential is zero along the z-axis, even though the magnetic field is non-zero?

The magnetic field is determined by how the magnetic potential changes according to $\vec{B} = \nabla \times \vec{A}$, not just by the potential value at a particular point. This is similar to a situation in classical mechanics where the gravitational potential energy of an object placed on the surface of the earth is defined to be zero, but the gravitational force is not zero. Similarly, zero acceleration does not necessarily imply zero velocity, either.

We can also approach this result of zero magnetic vector potential based on the symmetry of the problem. For any current segment on a ring, there will always be another segment with the current flowing in the opposite direction across the diameter. Therefore, the magnetic potentials from those two segments will cancel out.

We conclude this chapter by mentioning an electromagnetic "four-potential," which combines the electric scalar potential and the magnetic vector potential functions. It is a subject of more advanced field theories, where the electric and magnetic fields are considered as a unified field according to the full Helmholtz theorem.

## CHAPTER 9

Electromagnetic Induction

9.1 MOTIONAL EMF

Up until this chapter, we have dealt with static electric and magnetic fields from the charge and current sources that are fixed in space. Now, let's consider the following situation. A conductor of length L moves through a region of a constant magnetic field with speed v_x. We assume that the magnetic field is pointing out of the page along the $+z$ direction, and the motion is in the $+x$ direction. Although the following diagram does not show it, the two ends of the conductor are connected by a wire, forming a complete circuit with resistance R.

As the electrons within the conductor move under a constant magnetic field, magnetic force will be exerted on them. According to the Lorentz force formula, the magnetic force points in the $+y$ direction with force per unit charge of magnitude $v_x B$.

This force generates the movement of the electrons along the length of the conductor. In other words, there is *emf* or $\mathcal{E}$ induced between two ends of the conductor, which can be calculated by integrating this constant amount of force per unit charge over its length:

$$\mathcal{E} = v_x B L.$$

Also, with Ohm's law, the amount of this charge flow can be determined as $I = \frac{\mathcal{E}}{R} = \frac{v_x B L}{R}$. This expression of current tells us how fast the electrons

DOI: 10.1201/9781003397496-9

are moving along the $+y$ direction with speed v_y. With λ denoting the linear charge density within the conductor, $I = \frac{dq}{dt} = \frac{\lambda dy}{dt} = \lambda v_y$, or $v_y = \frac{I}{\lambda}$.

Due to this movement of charges with speed (v_y), the conductor itself is also subject to the magnetic force. According to the right-hand rule and keeping in mind that the electrons have negative charges, the direction of this magnetic force will be in the $-x$ direction, opposite to the motion of the conductor, which is moving in the $+x$ direction with speed (v_x). The amount of this force per unit charge is $v_y B$. The total force acting on the conductor can be obtained again by integrating the force on each moving charge along the entire length of the conductor.

$$F = \int_0^L (\lambda dy) v_y B = \lambda L \left(\frac{I}{\lambda} \right) B = LIB.$$

In other words, to move a conductor at constant velocity v_x inside of a region of the magnetic field B, one must counteract this magnetic force that opposes the motion. The amount of mechanical work per time, or power, that needs to be provided is:

$$P_{\text{mechanical}} = F v_x = LIB v_x.$$

It is precisely equal to the amount of electric power dissipated in the circuit:

$$P_{\text{electric}} = I\mathcal{E} = I v_x BL.$$

This result means that the mechanical work generates electrical energy. The laws of electromagnetism are consistent with the principle of energy conservation.

```
# Code Block 9.1

import numpy as np
import matplotlib.pyplot as plt
from matplotlib.patches import Rectangle

ax = plt.figure(figsize=(3,4)).gca()
ax.add_patch(Rectangle((-1,-1),2,2,color='#CCCCCC'))
ax.plot((0,0),(-0.75,0.75),linewidth=6,alpha=0.5,color='k')
```

```
step = 0.4
xgrid, ygrid = np.meshgrid(np.arange(-1,1+step,step),
                           np.arange(-1,1+step,step),
                           indexing='ij')
ax.scatter(xgrid,ygrid,marker='.',color='gray')
ax.axis('equal')
ax.axis('square')
ax.set_xlim((-1.2,3))
ax.set_ylim((-1.2,1.2))
ax.axis('off')

ax.quiver(1.5,-0.5,1,0,angles='xy',scale_units='xy',scale=1)
ax.quiver(1.5,-0.5,0,1,angles='xy',scale_units='xy',scale=1)
ax.quiver(+0.25,0,+1,0,angles='xy',scale_units='xy',scale=2)
ax.quiver(-0.25,0,-1,0,angles='xy',scale_units='xy',scale=2)
ax.text(2.25,-0.25,r'$v_x$',fontsize=fs)
ax.text(1.5,+0.75,r'$v_y$',fontsize=fs)
ax.text(+0.25,+0.3,r'$F_{mech}$',fontsize=fs)
ax.text(-0.75,-0.5,r'$F_{mag}$',fontsize=fs)
plt.tight_layout()
plt.savefig('fig_ch9_motional_emf.pdf')
plt.show()
```

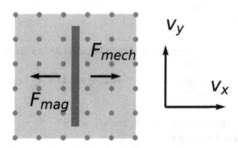

Figure 9.1

9.2 FARADAY'S LAW

In the following code block, we will present three different scenarios involving a square loop of a conducting wire and a region of a magnetic field. This region is shown as a gray square, just like the previous figure. Regularly spaced dots indicate that the magnetic field is pointing out of the page.

```
# Code Block 9.2

fig, axs = plt.subplots(3,1,figsize=(3,6),sharey=True)
fs = 12

for i in range(3):
    ax = axs[i]

    ax.add_patch(Rectangle((-1,-1),2,2,color='#CCCCCC'))
    ax.add_patch(Rectangle((0.5,-0.5),1,1,linewidth=6,alpha=0.5,
                           edgecolor='black',facecolor='white'))
    step = 0.25
    xgrid, ygrid = np.meshgrid(np.arange(-1,1+step,step),
                               np.arange(-1,1+step,step),
                               indexing='ij')
    ax.scatter(xgrid,ygrid,marker='.',color='gray')

    ax.quiver(0.3,0.5,0,-1,angles='xy',scale_units='xy',scale=1,
              linewidth=6)
    ax.text(0.05,0,r'$I$',fontsize=fs)

    ax.axis('equal')
    ax.axis('square')
    ax.set_xlim((-3,3))
    ax.set_ylim((-1.5,1.5))
    ax.axis('off')

axs[0].quiver(1.5,0,1,0,angles='xy',scale_units='xy',scale=1,
              linewidth=6)
axs[1].quiver(-1,0,-1,0,angles='xy',scale_units='xy',scale=1,
              linewidth=6)

axs[0].set_title("(a) Moving loop",loc='center',
                 fontsize=fs)
axs[1].set_title("(b) Moving magnetic field",loc='center',
                 fontsize=fs)
axs[2].set_title("(c) Decreasing magnetic field",loc='center',
                 fontsize=fs)

plt.tight_layout()
plt.savefig('fig_ch9_faraday_3cases.pdf')
plt.show()
```

(a) Moving loop

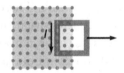

(b) Moving magnetic field

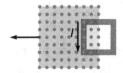

(c) Decreasing magnetic field

Figure 9.2

In the first case, the loop is pulled to the right. This is the same situation as the previous motional *emf* example. There will be induced *emf* on the segment of the loop lying within the magnetic field, and there will be a flow of electrons in the clockwise direction within the loop. Since the direction of electric current is defined conventionally as the direction of the flow of positive charges, the electric current is in the counterclockwise direction.

In the second case, the loop is fixed in space, but the region of the magnetic field moves to the left. Intuitively and unsurprisingly, this situation also produces a counterclockwise electric current in the square loop because the relative motion of the conducting loop and the region of the magnetic field is identical to the first case.

In the third case, nothing is moving, but the strength of the magnetic field is decreasing within the gray region. The magnetic field is no longer static, since it is changing over time. There are no moving charges relative to the magnetic field, yet this situation, as verified by experiments, produces a counterclockwise electric current, too. What is causing the charges to move? We need to extend our laws of electromagnetism, and this extension is given by Faraday's law.

According to Faraday's law,

$$\mathcal{E} = -\frac{d\Phi_B}{dt}.$$

$\mathcal{E}$ or *emf* equals the amount of work to push a unit charge around a closed path or a complete circuit and is defined as $\mathcal{E} = \oint \vec{E} \cdot d\vec{l}$. Φ_B denotes the magnetic flux, or the amount of magnetic field through a surface enclosed by the same loop for calculating $\mathcal{E}$, similar to how we defined electric flux. Mathematically, $\Phi_B = \iint \vec{B} \cdot d\vec{a}$.

In the three examples above, the square loop of side length L forms a closed path. The surface vector $d\vec{a}$ for calculating magnetic flux is parallel to the magnetic field $\vec{B}$. Let A be the shaded subarea of the square loop, where a non-zero, constant magnetic field is present. Then, the rest of the square loop area without any magnetic field is given by $L^2 - A$. The total magnetic flux through the square area is $\Phi_B = B \cdot A + 0 \cdot (L^2 - A) = BA$.

Let's examine the above three situations with Faraday's law. In the first and the second cases, the area A through which the magnetic field penetrates decreases because of the motion of either the loop or the region of the magnetic field. The rate of change in A is $\frac{dA}{dt} = -Lv_x$, where the negative sign indicates that the area is decreasing. Since the magnetic field is assumed to be constant in these two cases, the change in the magnetic flux is entirely due to the change in A, or

$$\frac{d\Phi_B}{dt} = B\frac{dA}{dt} = -BLv_x.$$

After applying the negative sign in Faraday's law, we obtain the identical expression for $\mathcal{E}$ as we did with the motional *emf* analysis.

In the motional *emf* analysis of the previous section, the right-hand rule for determining the direction of magnetic force revealed that the electric current flows in the counterclockwise direction. Can we also determine the direction of the induced current with Faraday's law? The answer is yes, and it is related to the negative sign in Faraday's law. Also known as Lenz's law, the negative sign signifies that the induced *emf* opposes the change in the magnetic flux. In all three cases, the magnetic flux through the square area decreases over time, and an additional magnetic field that points out of the page would oppose this change in magnetic flux. The counterclockwise current in the square loop would generate such an additional magnetic field. Hence, we conclude that the direction of the induced current is the counterclockwise direction in the square circuit.

The motional *emf* analysis is only applicable for the first case, but Faraday's law, with Lenz's law for determining the direction of the induced current, is a more general statement and can be applied to all three cases. In the third case, nothing is moving, but there is still a change in magnetic flux from the time-varying magnetic field. To oppose the decrease in magnetic flux, a counterclockwise current is similarly induced in the square circuit, which adds to the magnetic field pointing in the out-of-page direction. If the magnetic flux were increasing, the induced current would flow in the clockwise direction.

Let us rewrite Faraday's law as follows:

$$\frac{d\Phi_B}{dt} = \iint_{\text{Area}} \frac{\partial \vec{B}}{\partial t} \cdot d\vec{a} = -\oint_{\text{Contour}} \vec{E} \cdot d\vec{l} = -\iint_{\text{Area}} (\nabla \times \vec{E}) \cdot d\vec{a},$$

where we used Stokes' theorem $\iint_{\text{Area}} (\nabla \times \vec{K}) \cdot d\vec{a} = \oint_{\text{Contour}} \vec{K} \cdot d\vec{l}$. We have also made a careful distinction between the total time derivative of the flux and the partial derivative of the magnetic field inside the surface integral. The magnetic field $\vec{B}$ is a multivariate function of spatial and time variables, necessitating the use of partial derivatives to represent its change. In the case of magnetic flux, the spatial variables are integrated out over a specified area. As a result, Φ_B becomes solely dependent on time, making the total derivative appropriate. The mathematical rule for taking a differentiation of a multivariate function within an integral is known as Leibniz' rule.

By comparing the two surface integrals in the above expression, we obtain a differential form of Faraday's law:

$$\nabla \times \vec{E} = -\frac{\partial \vec{B}}{\partial t}.$$

This simple equation underlies how electricity is generated and profoundly impacts our everyday lives that rely on the use of electricity.

9.3 LC CIRCUIT

An inductor is another useful electronic circuit element whose function is based on electromagnetic induction phenomena. The simplest inductor is a coil of wire or a solenoid. The total magnetic field of the solenoid is the superposition of the magnetic field from each turn of the coil, and to a good approximation, it is confined within the interior of a solenoid with only a negligible magnetic field outside. Furthermore, the interior magnetic field is uniform and points along the axis of the solenoid. This situation is similar to that of a capacitor, where there is a uniform electric field between two parallel plates and a negligible electric field outside.

In the following code block, we calculate the magnetic field of a solenoid, `B_solenoid`, and demonstrate its uniformity inside. The magnetic field from each coil of a solenoid is calculated with the `currents_along_circle()` and `get_magnetic_field()` functions, which were used previously. We model a solenoid as a cylinder of length h, extending from $-\frac{h}{2}$ to $+\frac{h}{2}$ along the z-axis. Its circular cross-section has a radius of R and is centered at $x = 0$ and $y = 0$. The solenoid consists of N coils that are stacked on top of each other and separated by the distance $\frac{h}{N-1}$. The total magnetic field of a solenoid is the sum of the magnetic fields of all N coils, and it will be calculated at a few sample observation points along the x-axis.

```
# Code Block 9.3

# The following functions have already been defined
# and used in the earlier chapters.

# Excerpt from Code Block 5.3

def currents_along_circle (I,R,d_phi):
    # Create current distribution for a ring with radius R at origin.
    # phi is the angle from the positive x-axis.
    # p = (x,y,0)
    # curr = I*d_phi * (-y,x,0)
    phi = np.arange(0,2*np.pi+d_phi,d_phi)
    N = len(phi)
    p = np.zeros((3,N))
    x, y = R*np.cos(phi), R*np.sin(phi)
    p[0], p[1] = x, y

    curr = np.zeros((3,N))
    curr[0], curr[1] = -y/R, x/R
```

```
        curr = curr*I*R*d_phi
        return p, curr

# Excerpt from Code Block 5.5

def cross_product (A, B):
    # A and B: 3 by N arrays
    # This function is the same as np.cross(A.T,B.T).T
    V = np.zeros(A.shape)
    V[0] = A[1]*B[2] - A[2]*B[1]
    V[1] = A[2]*B[0] - A[0]*B[2]
    V[2] = A[0]*B[1] - A[1]*B[0]
    return V

def get_magnetic_field (p,p_curr,curr):

    # p: observation points
    # p_curr: position coordinate of current
    # curr: vector of current (times dl)

    assert p_curr.shape[1]==curr.shape[1]

    mu0 = 4*np.pi*(10**-7)
    M = p.shape[1]
    N = p_curr.shape[1]

    B = np.zeros(p.shape)

    for i in range(M):
        p_tmp = np.reshape(p[:,i],(3,1)).dot(np.ones((1,N)))
        r = p_tmp - p_curr # displacement vector: 3 by N
        dist3 = np.sqrt(np.sum(r**2,axis=0))**3 # vector of length N

        # Warning: dist3 could be very small or zero.
        # In such cases, the calculated B values will be unreliable.
        dB = cross_product(curr,r) / dist3

        # Cross product can be also calculated with a built-in
        # function in the numpy module, np.cross.
        #dB = np.cross(curr.T,r.T).T / dist3

        B[:,i] = np.nansum(dB,axis=1)
    B = (mu0/(4*np.pi))*B

    # Magnitude of B can be calculated by:
    #B_mag = np.sqrt(np.sum(B**2,axis=1))

    return B
```

```
# Excerpt from Code Block 5.8

def points_cartesian_xz (xmin=-1,xmax=1,zmin=-1,zmax=1,delta=0.1):
    x = np.arange(xmin,xmax+delta,delta)
    z = np.arange(zmin,zmax+delta,delta)
    y = 0
    xv,yv,zv = np.meshgrid(x,y,z,indexing='ij')
    p = np.vstack((xv.flatten(),yv.flatten(),zv.flatten()))
    return p
```

```
# Code Block 9.4

# Magnetic field of a solenoid.

I = 1 # current in solenoid
N = 100 # number of turns or coils
R = 0.5 # radius of each coil
h = 10 # length of solenoid

mu0 = 4*np.pi*(10**-7)

p = points_cartesian_xz(xmin=-1.1,xmax=1.1,zmin=0,zmax=0,delta=0.2)
B_solenoid = np.zeros(p.shape)

for i in range(N):
    p_curr, curr = currents_along_circle(I,R,np.pi/400)
    # The currents_along_circle() function
    # assumes that the circular coil is at z = 0,
    # so the z-position, p_curr[2], needs to be corrected
    # for each coil of the solenoid.
    p_curr[2] = i*h/(N-1)-h/2

    B_coil = get_magnetic_field(p,p_curr,curr)
    B_solenoid = B_solenoid + B_coil

# magnitude of total field
B_total = np.sqrt(np.sum(B_solenoid**2,axis=0))

fig = plt.figure(figsize=(3,3))
plt.scatter(p[0],B_total/mu0,color='#CCCCCC',label='Total $B$',s=90)
plt.scatter(p[0],B_solenoid[2]/mu0,color='#000000',label='$B_z$',s=10)
plt.xlabel('$x$')
plt.ylabel(r'$\frac{1}{\mu_0} B_{solenoid}$')
plt.xticks((-1,0,1))
plt.yticks((0,5,10))
plt.ylim((-2,16))
plt.legend(framealpha=1)
plt.tight_layout()
```

```
plt.savefig('fig_ch9_solenoid.pdf')
plt.show()
```

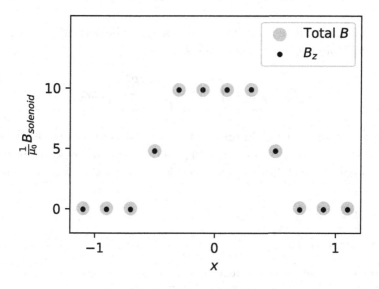

Figure 9.3

The above plot demonstrates that there is a uniform magnetic field inside the solenoid ($-0.5 < x < +0.5$) and a zero field outside ($x < -0.5$ and $x > +0.5$) when we take a slice across the x-axis at $y = z = 0$. The plot also indicates that the magnetic field points straight along the z-axis since the z-component of the magnetic field is equal to the magnitude of the total field, or $B_x = B_y = 0$.

We can calculate the magnetic field of this solenoid at different z values by adjusting the input arguments `zmin` and `zmax` of the `points_cartesian_xz()` function. As long as z is sufficiently smaller than h (in other words, in the limit of an infinitely long solenoid), we obtain the same result.

Let's delve into this further analytically. We can work with our earlier result about the magnetic field of a current ring, $\vec{B}_{\text{ring}}$, along its center axis. Consider a section of the solenoid with a thickness of dz at position z. There will be $\frac{N}{h}dz$ turns of coils, and the total amount of current through them will be $dI = I\frac{N}{h}dz$. The magnetic field at the middle of the solenoid, generated by this part of the solenoid, is

$$d\vec{B} = \frac{\mu_0 R^2}{2\left(R^2 + z^2\right)^{3/2}} dI\hat{z} = \frac{\mu_0 N I R^2}{2h\left(R^2 + z^2\right)^{3/2}} dz\hat{z}.$$

The total magnetic field within the solenoid can be determined by adding the fields from every section in the solenoid:

$$\vec{B}_{\text{solenoid}} = \int_{-h/2}^{+h/2} \frac{\mu_0 N I R^2}{2h\left(R^2 + z^2\right)^{3/2}} dz\hat{z}.$$

Because the magnetic field inside of a solenoid is uniform, the above result about the center of the solenoid generalizes to other positions within the solenoid. The following code block performs the integral symbolically.

```
# Code Block 9.5

import sympy as sym

# Magnetic field inside a solenoid.
R, h, z, mu, N, I = sym.symbols('R h z \mu_0 N I')
B_sol = sym.integrate(mu*N*I*R**2/(2*h*sym.sqrt(R**2+z**2)**3),
                      (z,-h/2,+h/2))
display(B_sol.nsimplify())
```

$$\frac{I N \mu_0}{2R \sqrt{1 + \frac{h^2}{4R^2}}}$$

For a long solenoid with $\frac{h}{2R} \gg 1$, the above analytic solution approaches $\vec{B}_{\text{solenoid}} = \frac{\mu_0 N I}{h}\hat{z}$. For $N = 100$, $h = 10$, and $I = 1$, we have $B_{\text{solenoid}} = 10\mu_0$, consistent with the result from the numerical calculation.

The magnetic field of a solenoid can also be determined with Ampere's law. Consider a rectangular Amperian loop of length h with one of its sides lying inside of the solenoid and its opposite side, outside. The total current enclosed by this rectangle is NI, and the line integral along the Ameprian loop is hB_{solenoid}, readily confirming that the magnitude of the solenoid's magnetic field is $\mu_0 N I / h$.

The magnetic flux of each coil in a solenoid is given by the product of the coil's circular area, πR^2, and the magnitude of the magnetic field, $\mu_0 NI/h$. If there is a temporal change in current, there will be an induced emf in each ring. The total induced emf of a solenoid that consists of N connected coils will be:

$$\mathcal{E} = -N\frac{d\Phi_B}{dt} = -\frac{\mu_0 N^2 \pi R^2}{h}\frac{dI}{dt}.$$

where Φ_B is the magnetic flux through each coil.

Lumping the product of constants as L, we can rewrite the above expression as:

$$\mathcal{E} = -L\frac{dI}{dt},$$

where L is called the inductance and has a unit of henry (H). The inductance describes the inductor's ability to produce an emf in response to an instantaneous change in electric current, and it depends mainly on its geometrical factors, such as N, R, and h.

As we did for an RC circuit, let us examine an LC circuit by connecting an inductor with inductance L and a capacitor with capacitance C in series. According to Kirchhoff's rule, the sum of potential changes around a closed circuit equals zero, so the following equation describes an LC circuit:

$$L\frac{dI}{dt} + \frac{q}{C} = 0.$$

Using $I = \frac{dq}{dt}$,

$$L\frac{d^2q}{dt^2} + \frac{q}{C} = 0.$$

This second-order differential equation has a solution represented by a linear combination of $\cos()$ and $\sin()$ functions. Suppose the initial condition of an LC circuit is such that the capacitor is charged with Q_0, and there is no current initially at $t = 0$. Then, the amount of charge on the capacitor is given by $q(t) = Q_0 \cos\left(\frac{t}{\sqrt{LC}}\right)$, and the current through the inductor is given by $I(t) = -\frac{Q_0}{\sqrt{LC}} \sin\left(\frac{t}{\sqrt{LC}}\right)$. The following code block illustrates how to solve a differential equation using the sympy module.

```
# Code Block 9.6

# Analyze an LC circuit.
```

```
# Define inductance, capacitance, time, and initial charge.
L, C, t, Q0 = sym.symbols('L C t Q_0', positive=True)

# Define charge as a function.
q = sym.symbols('q', cls=sym.Function)

# Define the ordinary differential equation.
diff_eqn = sym.Eq(L*q(t).diff(t,2)+q(t)/C,0)

# Solve the differential equation.
q_func = sym.dsolve(diff_eqn,
                    ics={q(0):Q0,sym.diff(q(t),t).subs(t,0):0},
                    simplify=True).rhs

# Solve for current.
I_func = sym.diff(q_func,t)

# Display the results.
print('charge at C, q(t)')
display(q_func.nsimplify())

print('current through L, I(t)')
display(I_func.nsimplify())
```

charge at C, q(t)

$$Q_0 \cos\left(\frac{t}{\sqrt{C}\,\sqrt{L}}\right)$$

current through L, I(t)

$$-\frac{Q_0 \sin\left(\frac{t}{\sqrt{C}\,\sqrt{L}}\right)}{\sqrt{C}\,\sqrt{L}}$$

The following code block illustrates how an *LC* circuit behaves over time. Once two ends of a charged capacitor are connected to an inductor, the capacitor starts discharging its charge through the inductor, and this discharge current generates a magnetic field inside the inductor. According to Faraday's law, an *emf* is induced in the inductor in response to this change of magnetic flux, opposing the current flow. As a result, the discharge process takes time. After a sufficient amount of time has passed, the capacitor is completely discharged, and even at this point, the current continues because the inductor, according to Faraday's law, opposes the change in magnetic flux. This leads to the recharging of the capacitor with opposite polarity, and then the discharging starts again. The discharging-charging process, accompanied by the switching of the polarity, has an oscillation period determined by the values of L and C, $f = \frac{1}{2\pi\sqrt{LC}}$. This LC circuit is analogous to a simple harmonic oscillator composed of a mass and a spring. The change-averse and inertia-like behavior of an inductor makes L analogous to mass m, and the potential-storing behavior of a capacitor makes C analogous to the reciprocal of spring constant k.

```python
# Code Block 9.7

# Visualizing an LC circuit with cartoon sketches.

def draw_LC_circuit(Q,q_now,I_now,t_now,V):

    # Plotting parameters.
    lw = 1 # linewidth.
    fs = 8 # fontsize.
    V_L, V_C = V

    # Positions of circuit elements: capacitor, inductor, wires.
    cap_x = [(-2,-2),(2,2)]
    cap_y = [(2,1),(2,1)]
    ind_x = [9.5,10.5,10.5,9.5]
    ind_y = [-1,-1,4,4]
    wire_x = [(0,0,0,10,10,0),(0,0,10,10,10,10)]
    wire_y = [(2,1,5,5,-1,-2),(5,-2,5,4,-2,-2)]

    plt.plot(cap_x, cap_y, color='gray', linewidth=lw)
    plt.fill(ind_x, ind_y, facecolor='none',
             edgecolor='gray', linewidth=lw)
    plt.plot(wire_x, wire_y, color='gray', linewidth=lw)
    plt.text(3,1,r'$C$',fontsize=fs)
    plt.text(8,1,r'$L$',fontsize=fs)

    # Number and positions of charges on each capacitor plate.
    q_num = int(q_now)
    x = np.linspace(-2,2,abs(q_num))
    y_t = 2.5*np.ones(len(x)) # location of top markers
```

```
    y_b = 0.5*np.ones(len(x)) # location of bottom markers
    if q_now >= 0:
        plt.scatter(x,y_t,s=10,marker='+',color='k',linewidths=lw)
        plt.scatter(x,y_b,s=10,marker='_',color='k',linewidths=lw)

    if q_now < 0:
        plt.scatter(x,y_t,s=10,marker='_',color='k',linewidths=lw)
        plt.scatter(x,y_b,s=10,marker='+',color='k',linewidths=lw)

    plt.quiver(5,+5,int(-I_now),0,scale_units='xy',scale=1)
    plt.quiver(5,-2,int(+I_now),0,scale_units='xy',scale=1)
    plt.yticks([])
    plt.xticks([])
    plt.axis('equal')
    plt.text(14,4,r"at t = %2.2f $\sqrt{LC}$"%(t_now),fontsize=fs)
    plt.text(14,2,r"$\Delta V$ at $L$ = %3.2f "%(V_L),fontsize=fs)
    plt.text(14,0,r"$\Delta V$ at $C$ = %3.2f "%(V_C),fontsize=fs)
    plt.box(on=False)

    return

# Simulate LC circuit at different times.
L_value, C_value = 1, 2
LC = L_value*C_value
Q_0 = 6 # initial number of charge on capacitor
t_range = np.array([0,0.25,0.5,0.75])*np.pi*np.sqrt(LC)

fig = plt.figure(figsize=(4,5))
for i, t_now in enumerate(t_range):
    # Calculate charge and current at t_now.
    value_dict = {Q0:Q_0, L:L_value, C:C_value, t:t_now}
    q_now = q_func.subs(value_dict).evalf()
    I_now = I_func.subs(value_dict).evalf()
    # time rate of change of current
    dI_now = I_func.diff(t).subs(value_dict).evalf()
    V_L, V_C = dI_now*L_value, q_now/C_value
    plt.subplot(len(t_range),1,i+1)
    draw_LC_circuit(Q_0, q_now, I_now, t_now/np.sqrt(LC), (V_L,V_C))

plt.tight_layout()
plt.savefig('fig_ch9_LC_circuit_cartoon.pdf')
plt.show()
```

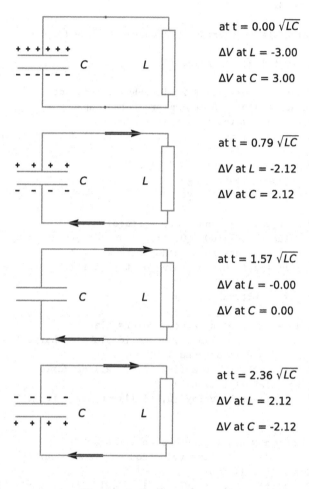

Figure 9.4

The following plot illustrates the oscillatory behavior of an *LC* circuit with a 3 *μF* capacitor and a 2 *μH* inductor in series. The capacitor is initially charged to 10 volts. An application of an *LC* circuit is to generate a periodic signal. The circuit can allow precise frequency control and tuning by selecting a specific combination of capacitance and inductance.

```
# Code Block 9.8

# Calculate the current and charge in L and C over a time interval.

fig = plt.figure(figsize=(3,3))

L_value = 2*10**(-6) # 2 micro-henry inductor
C_value = 3*10**(-6) # 3 micro-farad capacitor
Q_0 = 10 * C_value # initial charge
f = 1/(2*np.pi*np.sqrt(L_value*C_value)) # oscillation frequency

t_range = np.arange(0,(4+0.1)/f,0.01/f)
q = np.zeros(len(t_range))
I = np.zeros(len(t_range))

for i, t_value in enumerate(t_range):
    value_dict = {Q0:Q_0, L:L_value, C:C_value, t:t_value}
    q[i] = q_func.subs(value_dict).evalf()
    I[i] = I_func.subs(value_dict).evalf()

fig, ax1 = plt.subplots()

plot1, = ax1.plot(t_range,q,color='black',
                  linestyle='solid',label='charge (capacitor)')
ax1.set_ylabel('Coulombs')
ax1.set_yticks(np.array([-2,-1,0,1,2])*np.max(q))
ax1.set_xlabel('t (sec)')
ax1.set_xticks(np.array([0,0.5,1])*np.max(t_range))

ax2 = ax1.twinx()
plot2, = ax2.plot(t_range,I,color='black',
                  linestyle='dotted',label='current (inductor)')
ax2.set_ylabel('Amps')
ax2.set_yticks(np.array([-2,-1,0,1,2])*int(np.max(I)))

ax2.set_title('LC Circuit (L = %d $\mu$H, C = %d $\mu$F)\n'
              %(L_value*10**6,C_value*10**6))
ax2.legend(handles=[plot1, plot2],framealpha=1,loc='lower right')
plt.tight_layout()
plt.savefig('fig_ch9_LC_circuit.pdf')
plt.show()
```

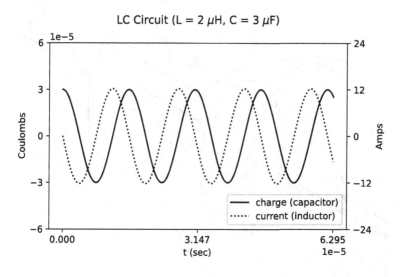

Figure 9.5

9.4   MAXWELL'S CORRECTION TO AMPERE'S LAW

The operation of an LC circuit can be described as continuous and gradual switching between the electric field in the capacitor and the magnetic field in the inductor, where a change in one field leads to a change in the other. Faraday's law clearly states that a changing magnetic field can induce an emf. Maxwell suspected that a changing electric field might similarly induce magnetic phenomena, but none of the known laws of electromagnetism addressed such a possibility when he started working on the theory of electromagnetism. He hypothesized that Ampere's law known up to that point, $\oint_{\text{Contour}} \vec{B} \cdot d\vec{l} = \mu_0 \iint_{\text{Area}} \vec{J} \cdot \hat{n} da$, might not be complete.

In the LC circuit, the switching between the electromagnetic fields is mediated by the current flow in the circuit, but what exactly happens between the two plates of the capacitor? The electrical charges do not jump across the plates, so there is no physical current. We would not describe what is happening in the capacitor with current density, $\vec{J}$.

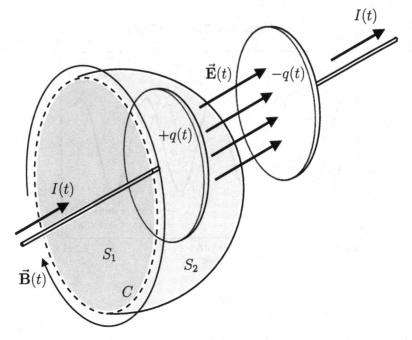

Figure 9.6

The above diagram shows a capacitor with two circular plates. The current flow $I(t)$ affects the amount of charge $q(t)$ on the plates, which is responsible for the electric field $E(t)$ between the plates. An Amperian loop, denoted as C, is drawn around the wire connected to the capacitor. Note C refers to the contour of the loop, not the capacitance. Let's consider two different surfaces that share the same boundary as this closed contour. The first surface, S_1, is the flat surface on the same plane as the loop, and the wire penetrates this surface. The second surface, S_2, is the half-sphere and goes between the two capacitor plates where no current passes.

Around the closed path C, there is a circulating magnetic field due to the current through the wire, according to Ampere's law. Then, according to Stokes' theorem, the line integral of this magnetic field around C can be expressed as a surface integral involving the curl of the magnetic field. This surface integral can be calculated with any surface as long as they have the same boundary. In particular, the two surfaces of S_1 and S_2 both encompass the contour C, so the surface integrals of the curl of the magnetic field through each surface should be equivalent.

For the surface S_1, which encloses the total current of $I(t)$, the line integral $\oint_C \vec{B}(t) \cdot \vec{dl}$ can be written as the surface integral $\iint_{S_1} (\nabla \times \vec{B}(t)) \cdot \hat{n} da$. This surface integral is also equal to the total sum of current inside times the free-space permeability, or $\mu_0 \iint_{S_1} \vec{J} \cdot \hat{n} da = \mu_0 I(t)$. However, for the second surface S_2, there is no current, so $\mu_0 \iint_{S_2} \vec{J} \cdot \hat{n} da = 0$.

This is puzzling, because the surface integral $\iint_{S_2} (\nabla \times \vec{B}(t)) \cdot \hat{n} da$ must be equal to the other surface integral $\iint_{S_1} (\nabla \times \vec{B}(t)) \cdot \hat{n} da$, which should be equal to $\oint_C \vec{B}(t) \cdot \vec{dl}$.

To resolve this apparent contradiction in the laws of electromagnetism, Maxwell proposed the concept of displacement current within the capacitor, which is different from the conduction current flowing through the wire in the circuit. Even though individual electrical charges do not jump across the gap between the two plates of a capacitor, there is a change in the amount of charges on the capacitor and an accompanying change in the electric field. The concept of displacement current accounts for these changes in the capacitor.

This relationship between $q(t)$ and $\vec{E}(t)$ can be derived from Gauss's law. Since the charge $q(t)$ on one capacitor plate is confined within the volume enclosed by S_1 and S_2,

$$\frac{q(t)}{\epsilon_0} = \iint_{S_1} \vec{0} \cdot \hat{n} da + \iint_{S_2} \vec{E}(t) \cdot \hat{n} da,$$

where the electric field is zero outside of the capacitor around S_1. Hence, the displacement current, defined as the rate of change of charge, can be defined in terms of the electric field:

$$\frac{dq(t)}{dt} = \epsilon_0 \iint_{S_2} \frac{\partial \vec{E}(t)}{\partial t} \cdot \hat{n} da,$$

where according to Leibniz' rule, a partial derivative is used inside the integral.

Thus, we present a corrected Ampere's law that includes both conduction current density $\vec{J}$ and time-varying electric field $\vec{E}(t)$ for any arbitrary contour and the surface area bounded by it:

$$\oint_{Contour} \vec{B}(t) \cdot \vec{dl} = \iint_{Area} (\nabla \times \vec{B}(t)) \cdot \hat{n} da$$

$$= \mu_0 \iint_{Area} \left(\vec{J} + \epsilon_0 \frac{\partial \vec{E}(t)}{\partial t} \right) \cdot \hat{n} da.$$

In a differential form, Ampere's law with Maxwell's correction states

$$\nabla \times \vec{B} = \mu_0 \vec{J} + \mu_0 \epsilon_0 \frac{\partial \vec{E}}{\partial t}.$$

Now, just like Faraday's law, Ampere's law with Maxwell's correction shows how the changing electric field gives rise to the magnetic field. In the next chapter, we will see such symmetric and mutual induction of electric and magnetic fields leads to electromagnetic waves.

## CHAPTER 10

Maxwell's Equations and Electromagnetic Wave

10.1 NO MAGNETIC MONOPOLE

Gauss's law, which we have studied earlier, states that the electric flux across a closed surface is proportional to the total enclosed charge, and its mathematical expression is: $\oiint_{\text{Area}} \vec{\mathbf{E}} \cdot d\vec{a} = \frac{q_{\text{enc}}}{\epsilon_0}$. This law expresses the relationship between electric charges and electric fields. We have also discussed that electrical charges come in two different types: positive and negative. The electric field and its flux point out of the positive charges, which act like a source, and flow into the negative charges, which act like a sink.

Analogous to the electrical charges, can we also consider a magnetic monopole, or an individual magnetic charge, that acts as a source or a sink of the magnetic field? Perhaps, are the two ends of a bar magnet magnetic monopoles of opposite polarity? Is the north end of the bar magnetic a "positive" magnetic charge, and the south end, a "negative" magnetic charge?

If a magnetic monopole exists, we expect there would be a version of Gauss's law for magnetism, which would state that the magnetic flux $\oiint_{\text{Area}} \vec{\mathbf{B}} \cdot d\vec{a}$ through a closed surface area is proportional to the total amount of magnetic charges enclosed within the volume bounded by this surface. According to the divergence theorem, this surface integral can be expressed as a volume integral of the divergence of the magnetic field:

$\oiint_{\text{Area}} \vec{B} \cdot d\vec{a} = \iiint_{\text{Volume}} (\nabla \cdot \vec{B}) dV$. However, we have already discussed that the magnetic field has zero divergence, or $\nabla \cdot \vec{B} = 0$. This was the reason why we have a vector potential for the magnetic field, and $\oiint_{\text{Area}} \vec{B} \cdot d\vec{a} = 0$, always.

The fact that $\oiint_{\text{Area}} \vec{B} \cdot d\vec{a} = 0$ or equivalently $\nabla \cdot \vec{B} = 0$ implies that no magnetic monopole can be found. Alternatively, if magnetic monopoles were to exist, they cannot be present in isolation by itself. A positive magnetic monopole must always be co-located with a negative magnetic monopole with equal magnitude, forming a magnetic dipole with a net-zero magnetic charge. Since the volume for the integral of Gauss's law can be chosen to be infinitesimally small, the magnetic dipole must exist as an inseparable pair with zero separation, and a single magnetic monopole can never be found by itself. Unlike the electric field, the source and sink of a magnetic field are located precisely at the same point in space, so the magnetic field always has zero divergence. Even if a magnet is broken into small pieces, there will never be a piece that carries a single magnetic pole. The north and south poles will always be perfectly matched.

10.2 MAXWELL'S EQUATIONS

Our study of electromagnetism has thus far unveiled the following observations and insights: Electric charges and currents underlie the formation of static electric and magnetic fields. There is a dynamic relationship between the electric and magnetic fields, where a change in one induces the other. Unlike electric charges, magnetic monopoles do not exist in isolation. These relations can be summarized in integral forms as follows:

$$\oiint_{\text{Area}} \vec{E} \cdot d\vec{a} = \frac{q_{\text{enc}}}{\epsilon_0} \quad \text{(Gauss's law)},$$

$$\oiint_{\text{Area}} \vec{B} \cdot d\vec{a} = 0 \quad \text{(No magnetic monopole)},$$

$$\oint_{\text{Contour}} \vec{E} \cdot d\vec{l} = -\frac{d}{dt} \iint_{\text{Area}} \vec{B} \cdot d\vec{a} \quad \text{(Faraday's law)},$$

$$\oint_{\text{Contour}} \vec{B} \cdot d\vec{l} = \mu_0 I_{\text{enc}} + \mu_0 \epsilon_0 \frac{d}{dt} \iint_{\text{Area}} \vec{E} \cdot d\vec{a}$$

$$\text{(Ampere's law with Maxwell's correction)}.$$

Collectively, these four equations are known as Maxwell's equations, and they, plus the Lorentz force equation $\vec{F} = q\vec{E} + q\vec{v} \times \vec{B}$, describe all of the fundamentals of electromagnetism. Maxwell's equations can also be expressed in differential forms:

$$\nabla \cdot \vec{E} = \frac{\rho}{\epsilon_0} \quad \text{(Gauss's law)},$$

$$\nabla \cdot \vec{B} = 0 \quad \text{(No magnetic monopole)},$$

$$\nabla \times \vec{E} = -\frac{\partial \vec{B}}{\partial t} \quad \text{(Faraday's law)},$$

$$\nabla \times \vec{B} = \mu_0 \vec{J} + \mu_0 \epsilon_0 \frac{\partial \vec{E}}{\partial t} \quad \text{(Ampere's law with Maxwell's correction)}.$$

In free space where there are no free charges and currents present (i.e., $\rho = 0$ and $\vec{J} = 0$), Maxwell's equations can be simplified as:

$$\nabla \cdot \vec{E} = 0$$
$$\nabla \cdot \vec{B} = 0$$
$$\nabla \times \vec{E} = -\frac{\partial \vec{B}}{\partial t}$$
$$\nabla \times \vec{B} = \mu_0 \epsilon_0 \frac{\partial \vec{E}}{\partial t}$$

10.3 WAVE EQUATION FROM MAXWELL'S EQUATIONS

Let's take a curl on Ampere's law in free space, or apply $\nabla \times$ to both sides of the fourth free-space Maxwell's equation. The left side of the equation becomes

$$\nabla \times (\nabla \times \vec{B}) = \nabla(\nabla \cdot \vec{B}) - \nabla^2 \vec{B} = \nabla(0) - \nabla^2 \vec{B} = -\nabla^2 \vec{B},$$

where we used the second derivative identity, $\nabla \times (\nabla \times \vec{K}) = \nabla(\nabla \cdot \vec{K}) - \nabla^2 \vec{K}$, from the chapter on vector calculus, combined with zero divergence of the magnetic field, which is the second Maxwell's equation.

The curl of the right side can be written as

$$\nabla \times (\mu_0 \epsilon_0 \frac{\partial \vec{E}}{\partial t}) = \mu_0 \epsilon_0 \frac{\partial}{\partial t}(\nabla \times \vec{E}) = \mu_0 \epsilon_0 \frac{\partial}{\partial t}(-\frac{\partial \vec{B}}{\partial t}) = -\mu_0 \epsilon_0 \frac{\partial^2 \vec{B}}{\partial t^2}.$$

Here we used the fact that the order of the spatial partial derivative in ∇ and the temporal derivative $\frac{\partial}{\partial t}$ can be swapped, or $\frac{\partial^2}{\partial x \partial t} = \frac{\partial^2}{\partial t \partial x}$ so that $\nabla \times \frac{\partial \vec{E}}{\partial t} = \frac{\partial}{\partial t}(\nabla \times \vec{E})$. Also, according to Faraday's law, the curl of the electric field can be expressed as the time derivative of the magnetic field.

By equating the left and right sides, we arrive at the following second-order partial differential equation for the magnetic field.

$$\nabla^2 \vec{B} = \mu_0 \epsilon_0 \frac{\partial^2 \vec{B}}{\partial t^2}.$$

We can also take a curl on Faraday's law, the third Maxwell's equation in free space. By taking into account the absence of electric charges in free space and Ampere's law, we arrive at another second-order partial differential equation for the electric field.

$$\nabla^2 \vec{E} = \mu_0 \epsilon_0 \frac{\partial^2 \vec{E}}{\partial t^2}.$$

The above results can be rewritten in the following general form:

$$\frac{\partial^2 \vec{U}}{\partial x^2} + \frac{\partial^2 \vec{U}}{\partial y^2} + \frac{\partial^2 \vec{U}}{\partial z^2} = \frac{1}{v^2} \frac{\partial^2 \vec{U}}{\partial t^2},$$

which is called a wave equation, and its solution propagates with speed v. This equation also appears in classical mechanics. For instance, when we deal with an acoustic wave in a solid medium with bulk modulus B and mass density ρ, the vector field $\vec{U}$ of the sound wave satisfies the same equation with speed $v = \sqrt{B/\rho}$.

Let's work with a simpler, one-dimensional wave equation with respect to the spatial variable z and time t:

$$\frac{\partial^2 U}{\partial z^2} = \frac{1}{v^2} \frac{\partial^2 U}{\partial t^2}.$$

This equation has a solution for any function $U(z, t)$ whose argument can be written as $(z-vt)$ or $(z+vt)$. In other words, if $U(z,t) = f(u) = f(z \pm vt)$, it satisfies the wave equation. We can verify this by taking the partial derivatives of $U(z, t)$.

$$\frac{\partial U}{\partial z} = \frac{df}{du} \frac{\partial u}{\partial z} = \frac{df}{du}.$$

$$\frac{\partial U}{\partial t} = \frac{df}{du} \frac{\partial u}{\partial t} = \pm v \frac{df}{du}.$$

Now, let's take the second derivatives.

$$\frac{\partial^2 U}{\partial z^2} = \frac{\partial}{\partial z} \left(\frac{df}{du} \right) = \frac{d^2 f}{du^2} \frac{\partial u}{\partial z} = \frac{d^2 f}{du^2}.$$

$$\frac{\partial^2 U}{\partial t^2} = \frac{\partial}{\partial t}\left(\pm v\frac{df}{du}\right) = \pm v\frac{d^2 f}{du^2}\frac{\partial u}{\partial t} = v^2\frac{d^2 f}{du^2}.$$

Hence, we have just shown that any function expressed as $f(z \pm vt)$ satisfies the one-dimensional wave equation. It is often common to express the solution as $f(z \pm vt) = U(kz \pm \omega t)$, where $\omega = vk$. The quantity k is known as the wave number, and ω is the angular frequency. Similarly, a solution to the three-dimensional wave equation would be expressed as $\vec{U}(\vec{k} \cdot \vec{r} \pm \omega t)$ with $\omega = \vec{v} \cdot \vec{k}$, where $\vec{v}$ is a three-dimensional velocity vector and $\vec{k}$ is a wave vector.

A function that can be written as $f(z-vt)$ would shift its snapshot profile $f(z)$ at $t = 0$ at constant speed v over time in the positive z direction. Likewise, $f(z+vt)$ will keep its shape and move in the opposite direction. The following code block illustrates how a packet of waves propagates in opposite directions: one moving to the right and the other to the left.

```python
# Code Block 10.1

import numpy as np
import matplotlib.pyplot as plt

def wave_packet (z,t,v=2,k=10):
    sig = 0.5
    f = z+v*t
    u = 0.5*np.exp(-f**2/2/sig**2)*np.sin(k*f)
    return u

fig = plt.figure(figsize=(3,3))
z = np.arange(-15,15,0.1)
t_range = np.arange(0,8,3)
c_range = np.linspace(0.8,0,len(t_range))
v = 2
for t,c in zip(t_range,c_range):
    plt.plot(z,wave_packet(z,t,v=+v)-1,color=(c,c,c),
            label='t = %d'%t)
    plt.plot(z,wave_packet(z,t,v=-v)+1,color=(c,c,c))
plt.legend(loc='center right')
plt.text(-13,1.1,'$f(z-vt)$')
plt.text(5,-0.9,'$f(z+vt)$')
plt.xlabel('z')
plt.xticks((-10,0,10))
plt.yticks(())
plt.title('Wave packet propagation (v=%d)'%v)
plt.tight_layout()
plt.savefig('fig_ch10_traveling_wave.pdf')
plt.show()
```

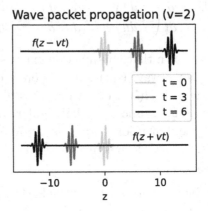

Figure 10.1

It is worth noting that a linear combination of the solutions to the same wave equation is also a solution. In other words, if $f_1(z, t)$ and $f_2(z, t)$ are individually a solution to the same wave equation, their linear combination, $c_1 f_1(z, t) \pm c_2 f_2(z, t)$ with arbitrary constants c_1 and c_2, is a solution, thanks to the linearity of the derivative operation.

As an interesting special case, consider $\sin(kz - \omega t)$ and $\sin(kz + \omega t)$ that describe two traveling sinusoidal waves moving in the opposite directions with the same speed $v = \frac{\omega}{k}$. Their sum is also a solution, and from the trigonometric identity, $\sin \alpha + \sin \beta = 2 \cos \frac{\alpha - \beta}{2} \sin \frac{\alpha + \beta}{2}$, this new solution is a standing wave, where the spatial variable z and the temporal variable t are separated.

$$\sin(kz - \omega t) + \sin(kz + \omega t) = 2 \cos(\omega t) \sin(kz).$$

```
# Code Block 10.2

# Standing waves

fig = plt.figure(figsize=(3,3))
z = np.arange(-10,10,0.1)
t_range = np.arange(-8,0,2)
c_range = np.linspace(0.8,0,len(t_range))

k = 1
w = 0.25
for t,c in zip(t_range,c_range):
    plt.subplot(3,1,1)
```

```
        plt.plot(z,np.sin(k*z-w*t),color=(c,c,c))
        plt.xlim((-10,10))
        plt.ylim((-1.1,1.1))
        plt.ylim((-2.1,2.1))
        plt.xticks(())
        plt.yticks(())
        plt.title(r'$\sin (kz-\omega t)$')

        plt.subplot(3,1,2)
        plt.plot(z,np.sin(k*z+w*t),color=(c,c,c))
        plt.xlim((-10,10))
        plt.ylim((-1.1,1.1))
        plt.ylim((-2.1,2.1))
        plt.xticks(())
        plt.yticks(())
        plt.title(r'$\sin (kz+\omega t)$')

        plt.subplot(3,1,3)
        plt.plot(z,np.sin(k*z-w*t)+np.sin(k*z+w*t),color=(c,c,c))
        plt.title(r'$\sin(kz-\omega t) + \sin(kz+\omega t)$')

# Curves with the same shade represent
# the waves at the same time.
plt.xlabel('z')
plt.xticks(())
plt.yticks(())
plt.xlim((-10,10))
plt.ylim((-2.1,2.1))
plt.tight_layout()
plt.savefig('fig_ch10_standing_wave.pdf')
plt.show()
```

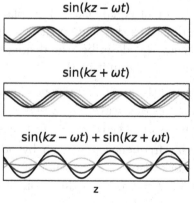

$\sin(kz - \omega t)$

$\sin(kz + \omega t)$

$\sin(kz - \omega t) + \sin(kz + \omega t)$

z

Figure 10.2

By examining the wave equations derived from Maxwell's equations, we conclude that the electromagnetic fields, $\vec{\mathbf{E}}$ and $\vec{\mathbf{B}}$, in free space are waves moving at speed given by $c \equiv \frac{1}{\sqrt{\epsilon_0 \mu_0}} = 2.99 \times 10^8$ m/s. Recognizing that this speed was, in fact, very close to the experimentally determined speed of light, Maxwell correctly proposed that light is an electromagnetic wave. This is one of the most celebrated triumphs of modern physics, unifying three previously distinct areas of physical science: electricity, magnetism, and optics.

10.4 MONOCHROMATIC PLANE WAVES

One class of functions that satisfy the free-space Maxwell's equations is a plane wave, whose wavefront, the set of points sharing the same position in their oscillation cycle, is a flat plane and moves perpendicular to this plane without bending or spreading out.

A plane wave is a good approximation for many forms of wave. For example, in classical optics, light rays coming from a far enough source onto a small enough cross-section (like the light from the Sun or a distant star that is being observed on the Earth) are effectively parallel to each other, so the perpendicular cross-sections of these rays can be treated as plane waves.

A sinusoidal plane wave at position $\vec{r}$ and time t can be mathematically represented by either $A\sin(\vec{k} \cdot \vec{r} - \omega t)$, $A\cos(\vec{k} \cdot \vec{r} - \omega t)$, or a complex exponential form $Ae^{i(\vec{k} \cdot \vec{r} - \omega t)}$. Here, A is a constant value denoting the maximum amplitude of the wave, and the argument $(\vec{k} \cdot \vec{r} - \omega t)$ is known as a phase or phase angle in wave mechanics. The wave vector $\vec{k}$ corresponds to the direction of wave propagation, and, hence, the wavefront of the same oscillatory cycle, or of equal phase, is a plane normal to $\vec{k}$. The angular frequency ω describes the angular displacement of the wave per unit time. The frequency, $\frac{\omega}{2\pi}$, describes the displacement of wave cycles per unit time, measured in the unit of Hz.

Both the wave vector and the angular frequency are related to the spatial and temporal periodicities of a wave. For a plane wave with wavelength λ and period T, moving at speed $v = \frac{\lambda}{T}$, the magnitude of the wave vector and the angular frequency are given by $|\vec{k}| = \frac{2\pi}{\lambda}$ and $\omega = \frac{2\pi}{T}$. The following code block displays a snapshot of a sinusoidal plane wave moving along a line of $y = x$ or in a direction of $\vec{k} = \frac{2\pi}{\lambda}(\frac{1}{\sqrt{2}}\hat{x} + \frac{1}{\sqrt{2}}\hat{y})$

at $t = 0$. The different gray levels correspond to different phases of the wave cycle.

```
# Code Block 10.3

# Visualize a sinusoidal plane wave, moving along a line of y=x.

A = 1 # amplitude
wavelength = 3
k = 2*np.pi/wavelength # wave vector

# a meshgrid for x, y, and z coordinates
N = 30
x, y, z = np.meshgrid(np.linspace(0,20,N),
                      np.linspace(0,20,N),
                      np.linspace(-10,10,N))

# sinuosidal plane wave at t=0.
wave = A*np.cos(k*x/np.sqrt(2)+k*y/np.sqrt(2))

# Display only a portion of a wave.
wave[(y - x > 8) | (y - x < -8)] = np.nan
wave[(y + x < 8) | (y + x > 28)] = np.nan

fig = plt.figure(figsize=(3,3))
ax = fig.add_subplot(111, projection='3d')
ax.scatter(x.flatten(),y.flatten(),z.flatten(),
           c=wave.flatten(), cmap='gray')

# Indicate the direction of wave propagation with an arrow.
ax.quiver3D(15,15,0,1.5*k,1.5*k,0, color='black')

ax.set_xlabel('x')
ax.set_ylabel('y')
ax.set_zlabel('z')
ax.set_xticks([0,10,20])
ax.set_yticks([0,10,20])
ax.set_zticks([-10,0,10])
ax.view_init(20, 120)
ax.set_box_aspect(aspect=None, zoom=0.8)
plt.tight_layout()
plt.savefig('fig_ch10_plane_wave_diag.pdf')
plt.show()
```

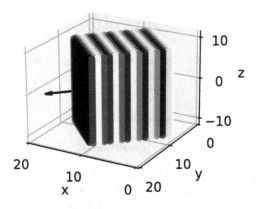

Figure 10.3

Now let's study a plane wave of the electromagnetic fields. We can start with a general expression for an electric field $\vec{E}(x, y, z, t)$:

$$\vec{E}(x, y, z, t) = E_x(x, y, z, t)\hat{x} + E_y(x, y, z, t)\hat{y} + E_z(x, y, z, t)\hat{z},$$

where $E_x(x, y, z, t)$ is an x-component of the electric field at position (x, y, z) and at time t. $E_y(x, y, z, t)$ and $E_z(x, y, z, t)$ correspond to the y and z components. As we have discussed in the previous section, the electric field in free space (i.e., no ρ and no $\vec{J}$) satisfies the wave equation, so we will consider an electric field that is a sinusoidal plane wave with a fixed angular frequency ω. Since the electromagnetic wave travels at speed c, we know its wavelength: $\lambda = \frac{2\pi c}{\omega}$. Such a wave with a single wavelength or a single frequency is referred to as a monochromatic wave. An example of a monochromatic electromagnetic wave is a single-colored laser.

In general, E_x, E_y, and E_z can be different from one another with distinct functional forms, but for a monochromatic sinusoidal plane wave,

$$E_x(x, y, z, t) = E_x(\vec{k} \cdot \vec{r} - \omega t) = \bar{E}_x e^{i(\vec{k} \cdot \vec{r} - \omega t)},$$

where $\bar{E}_x$ denotes a constant (potentially a complex number). Likewise, E_y and E_z are also functions of $\vec{k} \cdot \vec{r} - \omega t$: $E_y(x, y, z, t) = \bar{E}_y e^{i(\vec{k} \cdot \vec{r} - \omega t)}$ and $E_z(x, y, z, t) = \bar{E}_z e^{i(\vec{k} \cdot \vec{r} - \omega t)}$.

For simplicity, we assume that the wave is propagating along the z-axis with a wave vector $\vec{\mathbf{k}} = k\hat{\mathbf{z}}$. That means that the resulting electric field $\vec{\mathbf{E}}(x, y, z, t)$ can be expressed as

$$\vec{\mathbf{E}}(x, y, z, t) = \bar{E}_x e^{i(kz-\omega t)}\hat{\mathbf{x}} + \bar{E}_y e^{i(kz-\omega t)}\hat{\mathbf{y}} + \bar{E}_z e^{i(kz-\omega t)}\hat{\mathbf{z}}.$$

E_x, E_y, and E_z are the functions of $(kz - \omega t)$, independent of the variables x and y that are transverse coordinates to the wave propagation. At any fixed value of z at a specific time t, all the points lying on the xy-plane orthogonal to the z-axis have the same phase of the wave, as expected of a plane wave.

We can similarly express the magnetic field as:

$$\vec{\mathbf{B}}(x, y, z, t) = \bar{B}_x e^{i(kz-\omega t)}\hat{\mathbf{x}} + \bar{B}_y e^{i(kz-\omega t)}\hat{\mathbf{y}} + \bar{B}_z e^{i(kz-\omega t)}\hat{\mathbf{z}}.$$

Maxwell's equations further constrain $\bar{E}$'s and $\bar{B}$'s. In particular, because $\nabla \cdot \vec{\mathbf{E}} = \frac{\partial}{\partial x}E_x + \frac{\partial}{\partial y}E_y + \frac{\partial}{\partial z}E_z = 0$ in free space,

$$\frac{\partial}{\partial x}\left(\bar{E}_x e^{i(kz-\omega t)}\right) + \frac{\partial}{\partial y}\left(\bar{E}_y e^{i(kz-\omega t)}\right) + \frac{\partial}{\partial z}\left(\bar{E}_z e^{i(kz-\omega t)}\right)$$
$$= 0 + 0 + \frac{\partial}{\partial z}\left(\bar{E}_z e^{i(kz-\omega t)}\right)$$
$$= 0,$$

where the first two terms with partial derivatives with respect to x and y are trivially equal to zero because there is no dependence on x or y. For the above equation to hold true, the third term must be equal to zero, leading to the conclusion: $\bar{E}_z = 0$. Similarly, because $\nabla \cdot \vec{\mathbf{B}} = 0$, we also conclude that $\bar{B}_z = 0$. In other words, there is no z-component in the electromagnetic plane wave that propagates in the z direction. Hence, it is a purely transverse wave without any longitudinal component.

Now, let's apply Faraday's law, $\nabla \times \vec{\mathbf{E}} = -\frac{\partial \vec{\mathbf{B}}}{\partial t}$, and collect x and y components separately. A lot of terms such as $\frac{\partial}{\partial y}E_z$ or $\frac{\partial}{\partial x}E_z$ will disappear trivially, because $\bar{E}_z = \bar{B}_z = 0$.

The x components of the left and right sides of Faraday's law should be equal, or

$$-\frac{\partial}{\partial z}\left(\bar{E}_y e^{i(kz-\omega t)}\right) = -\frac{\partial}{\partial t}\left(\bar{B}_x e^{i(kz-\omega t)}\right)$$
$$-(ik)\bar{E}_y = (i\omega)\bar{B}_x.$$

The y components should also be equal to each other, or

$$\frac{\partial}{\partial z}\left(\bar{E}_x e^{i(kz-\omega t)}\right) = -\frac{\partial}{\partial t}\left(\bar{B}_y e^{i(kz-\omega t)}\right)$$

$$(ik)\bar{E}_x = (i\omega)\bar{B}_y.$$

These two relationships can be captured compactly by the following expression, using $\vec{\mathbf{k}} = k\hat{z}$:

$$\vec{\mathbf{B}} = \left(\frac{k}{\omega}\right)\hat{z} \times \vec{\mathbf{E}} = \frac{1}{\omega}\vec{\mathbf{k}} \times \vec{\mathbf{E}}.$$

This implies that the electric and magnetic fields of a monochromatic plane wave are perpendicular to each other. They also have the same phase, or are in-phase with each other, reaching their maximum or minimum values simultaneously.

By taking into account the mutual orthogonality of the three vectors ($\vec{\mathbf{E}}$, $\vec{\mathbf{B}}$, and $\vec{\mathbf{k}}$) as well as the fact that the speed of electromagnetic wave c is equal to ω/k, we can also conclude that:

$$\frac{|\vec{\mathbf{E}}|}{|\vec{\mathbf{B}}|} = \frac{\omega}{|\vec{\mathbf{k}}|} = c.$$

Monochromatic electromagnetic plane waves are sometimes illustrated with two in-phase, orthogonal sinusoidal functions, as shown in the following figure. The direction of wave propagation is along the z-axis, and the electromagnetic fields are uniform across each slice of the xy-plane. The electric field points along the x-axis, and the magnetic field is along the y-axis.

```python
# Code Block 10.4

def tidy_axis(ax,lim=3,maxz=10,elev=15,azim=20):
    # Clean up the axis.
    ax.set_xlim((-lim,lim))
    ax.set_ylim((-lim,lim))
    ax.set_zlim((0,maxz))
    ax.set_xticks((-1,0,1))
    ax.set_yticks((-1,0,1))
    ax.set_zticks((0,int(maxz/2),int(maxz)))
    ax.set_xlabel('x')
    ax.set_ylabel('y')
    ax.set_zlabel('z')
```

```
     ax.view_init(elev=elev,azim=azim,vertical_axis="x")

t = 0
k, w = 1, 1

maxz = 4*np.pi/k
z = np.arange(0,maxz+0.1,0.1)
Ex = np.sin(k*z-w*t)
By = np.sin(k*z-w*t)

fig = plt.figure(figsize=(3,3))
ax = fig.add_subplot(projection='3d')

ax.plot3D(Ex,np.zeros(len(z)),z,color='black')
ax.plot3D(np.zeros(len(z)),By,z,color='gray')
ax.plot3D([0,0],[0,0],[0,maxz],color='gray',linestyle='dashed')
tidy_axis(ax,lim=2,maxz=maxz)
plt.savefig('fig_ch10_plane_EMwave.pdf',bbox_inches='tight')
plt.show()
```

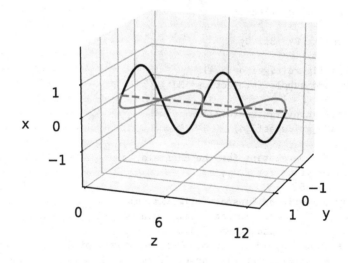

Figure 10.4

A different, perhaps more illustrative, visualization is possible, where multiple snapshots of electromagnetic fields are taken at sample slices of the xy plane at different times, as shown below.

```
# Code Block 10.5

example = 1 # Specify plane wave example.

def plane_wave(k,z,w,t,example=1):
    if example==1:
        Ex = np.sin(k*z-w*t)
        Ey = np.zeros(z.shape)
        Bx = np.zeros(z.shape)
        By = np.sin(k*z-w*t)
    if example==2:
        Ex = np.sin(k*z-w*t)
        Ey = np.sin(k*z-w*t)
        Bx = -np.sin(k*z-w*t)
        By = np.sin(k*z-w*t)
    if example==3:
        Ex = np.sin(k*z-w*t)
        Ey = np.sin(k*z-w*t-(np.pi/2))
        Bx = -np.sin(k*z-w*t-(np.pi/2))
        By = np.sin(k*z-w*t)
    if example==4:
        Ex = np.sin(k*z-w*t)
        Ey = np.sin(k*z-w*t+(np.pi/2))
        Bx = -np.sin(k*z-w*t+(np.pi/2))
        By = np.sin(k*z-w*t)

    return Ex, Ey, Bx, By

fig0 = plt.figure(figsize=(6,4))
ax = fig0.add_subplot(projection='3d')

fig = plt.figure(figsize=(8,8))
gs = fig.add_gridspec(3,4)

k, w = 1, 1 # parameters for the EM field.
# Set up the grids for x, y, z.
lim, num = 2, 5
x, y = np.meshgrid(np.linspace(-lim,lim,num),
                   np.linspace(-lim,lim,num),
                   indexing='ij')
z_range = (np.array([0,0.25,0.5,0.75]) + 1.0)*np.pi/k
t_range = np.array([0,np.pi/4,np.pi/2])/w
maxz = 4*np.pi/k

for i in range(4): # Update each subplot as we vary z.
    for j in range(3): # Update each subplot as we vary t.
        z = z_range[i]
        t = t_range[j]
        Ex, Ey, Bx, By = plane_wave(k,z,w,t,example=example)
```

```
        ax_sub = fig.add_subplot(gs[j,i])
        ax_sub.quiver(x,y,Ex,Ey,label='E',color='black',
                      angles='xy',scale_units='xy',scale=2)
        ax_sub.quiver(x,y,Bx,By,label='B',color='gray',
                      angles='xy',scale_units='xy',scale=2)
        ax_sub.set_title('(z, t) = (%2.1f, %2.1f)'%(z,t))

        ax_sub.set_xlim((-lim,lim))
        ax_sub.set_ylim((-lim,lim))
        ax_sub.set_xticks(())
        ax_sub.set_yticks(())
        ax_sub.set_xlabel('x')
        ax_sub.set_ylabel('y')
        ax_sub.axis('square')
        if (i==0) & (j==0):
            ax_sub.legend(loc='upper right')

        # Put an xy-plane at the corresponding z.
        z_grid = z+np.zeros(x.shape)
        ax.plot_surface(x,y,z_grid,color='#CCCCCC',alpha=0.1)

elev, azim = 15, 20 # Default viewing angle of the 3D plot.
if example == 3 or example == 4:
    # For the circular polarization visualizations,
    # use a longer range.
    maxz = maxz * 2
    # Use a different viewing angle.
    elev, azim = 5, 60

# Display the sinusoidal wave at t=0 in the top figure.
t = 0
z = np.arange(0,maxz/2+0.1,0.1)
Ex, Ey, Bx, By = plane_wave(k,z,w,t,example=example)
ax.plot3D(Ex,Ey,z,color='black')
ax.plot3D(Bx,By,z,color='gray')
ax.plot3D([0,0],[0,0],[0,maxz/2],color='gray',linestyle='dashed')
tidy_axis(ax,lim,maxz/2,elev=elev,azim=azim)

fig.tight_layout()
fig0.savefig('fig_ch10_plane_wave_ex%d.pdf'%example)
fig.savefig('fig_ch10_plane_wave_snapshots_ex%d.pdf'%example)
plt.show()
```

As we examine the snapshots of the electromagnetic fields, we see that the electric and magnetic fields are orthogonal to each other and oscillate with the same phase in time (the rows of snapshot plots) and in space (the columns).

It is, of course, possible for the axes of the electromagnetic fields to be at an angle with respect to the x or y axis. For instance, the wave

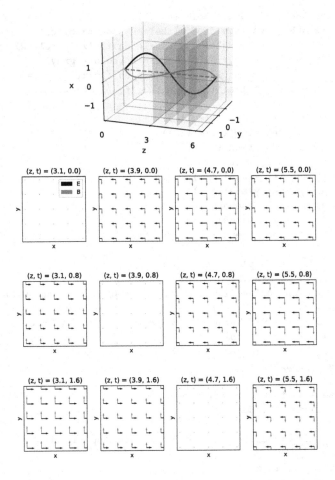

Figure 10.5

traveling along the z-axis may have non-zero components of E_x, E_y, B_x, and B_y, while $E_z = B_z = 0$, as visualized by the following figure. This plot corresponds to example=2 for the plane_wave() function. A different combination of E_x, E_y, E_z, B_x, B_y, and B_z would produce an electromagnetic wave propagating in different direction $\vec{\mathbf{k}}$, but in all cases, the relation of $\vec{\mathbf{B}} = \frac{1}{\omega}\vec{\mathbf{k}} \times \vec{\mathbf{E}}$ would still hold. These plane waves, where both $\vec{\mathbf{E}}$ and $\vec{\mathbf{B}}$ oscillate along two mutually perpendicular directions, are called linearly polarized. The light from the Sun is not polarized and does not have a preferred direction of oscillation, but when the sunlight reflects from the surface of water, it becomes linearly polarized. The interaction between electromagnetic fields and matter, which we did not get to explore much in this book, is an important and interesting topic.

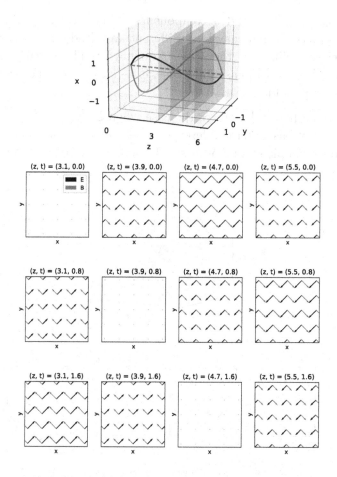

Figure 10.6

Other (non-linear) types of polarizations are possible, and we can analyze them by introducing complex amplitudes for the electromagnetic fields. A complex number can be expressed in a cartesian form, $a + ib$, or in a polar or exponential form, $\alpha e^{i\beta}$. They are related by: $\alpha = \sqrt{a^2 + b^2}$ and $\beta = \tan^{-1}(b/a)$.

We can rewrite the electric field as:

$$\vec{\mathbf{E}}(x, y, z, t) = |\bar{E}_x| e^{i\phi_x} e^{i(kz-\omega t)} \hat{\mathbf{x}} + |\bar{E}_y| e^{i\phi_y} e^{i(kz-\omega t)} \hat{\mathbf{y}}$$
$$= |\bar{E}_x| e^{i(kz-\omega t+\phi_x)} \hat{\mathbf{x}} + |\bar{E}_y| e^{i(kz-\omega t+\phi_y)} \hat{\mathbf{y}},$$

where the amplitudes of the electric field, $\bar{E}_x$ and $\bar{E}_y$, have been written in an exponential form with $|\bar{E}_x|$, $|\bar{E}_y|$, ϕ_x, and ϕ_y.

As a reminder, the x and y components of the electric field are independent functions that may have different amplitudes and phases, even though they oscillate at the same frequency ω. The linearly polarized electric field has zero phase difference, $\phi_x - \phi_y = 0$, but non-zero phase difference is possible.

The following figure illustrates the case when E_x and E_y have the same magnitude but has a phase difference of $+\frac{\pi}{2}$ (i.e., $|\bar{E}_x| = |\bar{E}_y|$ and $\phi_x - \phi_y = +\frac{\pi}{2}$). This example can be produced by setting `example=3` in the above code block.

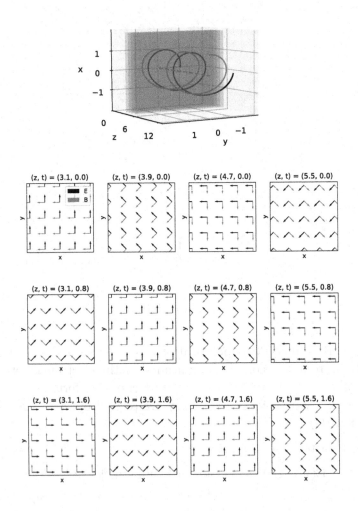

Figure 10.7

As seen from the series of temporal snapshots of the electric and magnetic fields, the phase difference in the complex amplitude produces a rotation of the electromagnetic fields, which is called circular polarization. If the phase difference is $-\frac{\pi}{2}$ (with example=4 in the code block), the field rotates in the opposite direction, as shown below. If the magnitudes of the two components are different, in addition to having a relative phase difference, the electromagnetic wave will exhibit elliptical polarization.

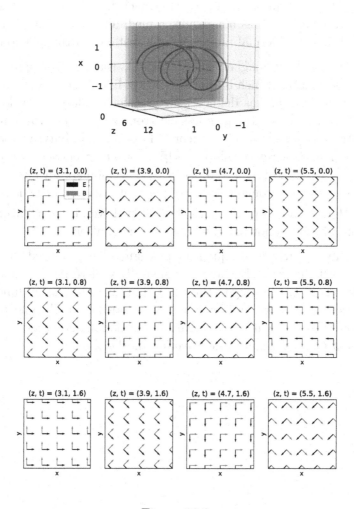

Figure 10.8

The propagation of the electromagnetic wave involves the flow or transmission of electromagnetic energy, which is quantified by a Poynting vector:

$$\vec{S} = \frac{1}{\mu_0}\vec{E} \times \vec{B}.$$

The Poynting vector points in the same direction as the wave vector, and its magnitude is the amount of energy transported across a unit area per unit of time. For mechanical waves like sound, water, or seismic waves, the energy of a wave can be intuitively imagined as being transmitted and exchanged through the motion and elasticity of the media. On the other hand, electromagnetic waves can travel through a vacuum, so their energy transmission does not require a medium, which is itself a rather remarkable fact. Thermal radiation is an instance of this phenomenon.

The theory of electromagnetism also suggests that electromagnetic wave carries momentum, which is another extraordinary fact. In classical mechanics, momentum was defined as a product of mass and velocity, or $\vec{p} = m\vec{v}$, so how could the massless electromagnetic wave carry momentum? According to Einstein's theory of special relativity, we have the following invariant relationship $E^2 - |\vec{p}|^2 c^2 = (mc^2)^2$ between momentum $\vec{p}$ and energy E, so for $m = 0$, $|\vec{p}| = \frac{E}{c}$. The momentum flux, the momentum of the electromagnetic wave per unit area and time, is given by $\vec{g} = \frac{\vec{S}}{c}$. Hence, the propulsion of a spaceship driven by momentum from the wind of light is not just a whimsical imagination.

Starting in the early 1900s, quantum mechanics has produced another intriguing, yet theoretically and experimentally well-verified, idea of wave-particle duality. According to this idea, light, as well as other fundamental particles, must be considered as having both wave-like and particle-like properties. We will have to save this discussion and other interesting aspects of the electromagnetic wave for another book.

Appendix

APPENDIX A: GETTING STARTED WITH PYTHON

Perhaps the most challenging step in following the codes in this book may be the first step of getting started with Python. Fortunately, there are a few user-friendly options at this point of writing.

The first option is a free, cloud-based Python environment like Google Colaboratory (or Colab) (`research.google.com/colaboratory`). You can open, edit, and run Python codes on a Jupyter Notebook environment using a browser. The second option is to download and install a distribution of Python that already includes relevant packages, such as `numpy` and `matplotlib`, and other valuable tools, such as Jupyter Notebook (`jupyter.org`). We recommend Anaconda Distribution (`www.anaconda.com`), which supports different operating systems (Windows, iOS, and Linux) and makes it easy to configure your computer. The third option is to install each module and dependency separately.

APPENDIX B: PYTHON PROGRAMMING BASICS

Whether it is Python or other programming languages, there are many standard practices, notations, structures, and techniques. This appendix goes over a few basic ideas if you are new to programming.

The following code block demonstrates the practice of using variables to hold values and do calculations.

```
# Code Block Appendix B.1

x = 5
y = 2
print(x+y)
print(x-y)
print(x*y)
```

```
print(x/y)
print(x**y)
```

```
7
3
10
2.5
25
```

Two powerful control structures are the if-conditionals and for-loops, as demonstrated below. The for-loop allows you to iterate a block of codes marked by indentation within the loop. The number of iterations is often specified with a built-in function range(). You can also easily work with individual elements in an array. If a condition given in the if-conditional is evaluated to be true, a set of codes marked by indentation will be executed.

```
# Code Block Appendix B.2

for i in range(5):
    print(i**2)
    if (i**2 == 9):
        print('This was a nice number.')
```

```
0
1
4
9
This was a nice number.
16
```

Another powerful practice in programming is to split up a complex task or procedure into smaller and more manageable chunks, which are called functions or modules. For example, you may be tasked to calculate an average of multiple values repeatedly. Then, it would be desirable to create a function that takes an arbitrary array of values as an input argument and returns its average.

In addition to being able to write your own functions or modules, it is also essential to be able to use well-written and widely adopted modules. For example, many common and critical computational routines are already written into the modules like numpy and scipy. By using them, instead of writing your own, your codes will be more readable and usable by others and will likely be more robust and less susceptible to errors.

```
# Code Block Appendix B.3

import numpy as np

def calculate_average(x):
    avg = 0
    for val in x:
        avg = avg + val
    return avg/len(x)

x = [1,5,3,7,2]

# Using a function created above.
print(calculate_average(x))

# Using a function from numpy module.
print(np.mean(np.array(x)))
```

3.6
3.6

Another important aspect of coding is to make mistakes and learn from them. The following code blocks demonstrate a few common error messages you might see.

```
# Code Block Appendix B.4

# Using numpy module without importing it ahead will
# generate an error message like:
# "NameError: name 'np' is not defined"

# Because we are demonstrating the importance of import,
# let's unimport or del numpy.
del numpy

x = np.array([1,2,3])
```

```
-------------------------------------------------------------
NameError                        Traceback (most recent call last)
<ipython-input-4-976cd16bfa8a> in <module>
      7 # Because we are demonstrating the importance of import,
      8 # let's unimport or del numpy.
----> 9 del numpy
     10
     11 x = np.array([1,2,3])

NameError: name 'numpy' is not defined
```

```
# Code Block Appendix B.5

# Python's indexing convention is to start at zero.
# The first element in an array is indexed by 0.
# The last element in an array is indexed by -1 or its length-1.
# If you try to index after the last element,
# you will get an error like:
# "IndexError: list index out of range"

x = [10,20,30]
print(x[0])
print(x[1])
print(x[2])
print(x[3])
```

```
10
20
30
-----------------------------------------------------------------
IndexError                      Traceback (most recent call last)
<ipython-input-5-1487e342efb9> in <module>
     11 print(x[1])
     12 print(x[2])
---> 13 print(x[3])

IndexError: list index out of range
```

APPENDIX C: CURVED COORDINATES

A Cartesian coordinate system allows us to specify a position in a three-dimensional space using a set of real numbers along orthogonal lines, namely (x, y, z). With a spherical coordinate system, this position can be specified with a different set of real numbers, (r, θ, ϕ). Consider a line segment drawn between the position and the origin. The radial distance r corresponds to the length of this line segment. The azimuthal angle θ is the angle between the line segment and the z-axis, and it takes on a value between 0 and π. The polar angle ϕ is the angle between the x-axis and the projection of the line segment on the xy-plane, and it takes on a value between 0 and 2π. The relationship between (x, y, z) and (r, θ, ϕ) is described by the following equalities:

$$x = r \sin \theta \cos \phi$$
$$y = r \sin \theta \sin \phi$$
$$z = r \cos \theta.$$

We can also define a set of inverse relationships.

$$r = \sqrt{x^2 + y^2 + z^2}$$
$$\theta = \arccos \frac{z}{r}$$
$$\phi = \arctan \frac{y}{x}.$$

The following code block illustrates the relationship between the Cartesian and spherical coordinates.

```
# Code Block Appendix C.1

import numpy as np
import matplotlib.pyplot as plt

def tidy_axis_3d(ax):
    # Clean up axes.
    ax.set_xlim((-1.1,1.1))
    ax.set_ylim((-1.1,1.1))
    ax.set_zlim((-1.1,1.1))
    ax.set_xticks(())
    ax.set_yticks(())
    ax.set_zticks(())
    ax.set_xlabel('x',labelpad=-15)
    ax.set_ylabel('y',labelpad=-15)
    ax.set_zlabel('z',labelpad=-15)

    # Draw the x, y, and z axes.
    lim = 1
    ax.plot3D([-lim,lim],[0,0],[0,0],color='gray',linewidth=0.5)
    ax.plot3D([0,0],[-lim,lim],[0,0],color='gray',linewidth=0.5)
    ax.plot3D([0,0],[0,0],[-lim,lim],color='gray',linewidth=0.5)
    ax.set_box_aspect([1,1,1])
    return

# Spherical coordinates
r = 1
```

```
step = np.pi/25
theta, phi = np.mgrid[0:np.pi+step:step, 0:2*np.pi+step:step]
x = r*np.sin(theta)*np.cos(phi)
y = r*np.sin(theta)*np.sin(phi)
z = r*np.cos(theta)

fig = plt.figure(figsize=(3,3))
ax = fig.add_subplot(projection='3d')
ax.plot_surface(x,y,z, cmap='gray')
tidy_axis_3d(ax)
plt.savefig('fig_appC_sphere.pdf', bbox_inches="tight")
plt.show()
```

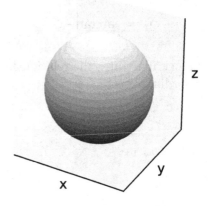

Figure S1

The following plots illustrate the range of space specified by θ and ϕ at the same distance from the origin (i.e., $r = 1$).

```
# Code Block Appendix C.2

# Visualize the theta and phi coordinates.

r = 1
step = np.pi/25
nrow = 5

fig = plt.figure(figsize=(4,10))
subfigs = fig.subfigures(nrow,1)
for i in range(nrow):
    subfigs[i].suptitle(r'$0 \leq \phi < \frac{%d}{5} 2\pi$'%(i+1))
```

```python
    ax = subfigs[i].add_subplot(1,2,1,projection='3d')
    theta, phi = np.mgrid[np.pi/5:np.pi/2+step:step,
                          0:(i+1)*step*10+step:step]
    x = r*np.sin(theta)*np.cos(phi)
    y = r*np.sin(theta)*np.sin(phi)
    z = r*np.cos(theta)
    ax.plot_surface(x,y,z, cmap='gray')
    tidy_axis_3d(ax)

    ax = subfigs[i].add_subplot(1,2,2,projection='3d')
    ax.plot_wireframe(x,y,z, color='gray')
    tidy_axis_3d(ax)
    ax.set_zlabel('')
    ax.view_init(90,-90)

plt.tight_layout()
plt.savefig("fig_appC_spherical_phi.pdf", bbox_inches="tight")
plt.show()

fig = plt.figure(figsize=(4,10))
subfigs = fig.subfigures(nrow,1)
for i in range(nrow):
    subfigs[i].suptitle(r'$0 \leq \theta < \frac{%d}{5} \pi$'%(i+1))

    ax = subfigs[i].add_subplot(1,2,1,projection='3d')
    theta, phi = np.mgrid[0:(i+1)*step*5+step:step,
                          0:2*np.pi+step:step]
    x = r*np.sin(theta)*np.cos(phi)
    y = r*np.sin(theta)*np.sin(phi)
    z = r*np.cos(theta)
    ax.plot_surface(x,y,z, cmap='gray')
    tidy_axis_3d(ax)

    ax = subfigs[i].add_subplot(1,2,2,projection='3d')
    ax.plot_wireframe(x,y,z, color='gray', linewidth=0.5)
    tidy_axis_3d(ax)
    ax.set_ylabel('')
    ax.view_init(0,90)

plt.tight_layout()
plt.savefig("fig_appC_spherical_theta.pdf", bbox_inches="tight")
plt.show()
```

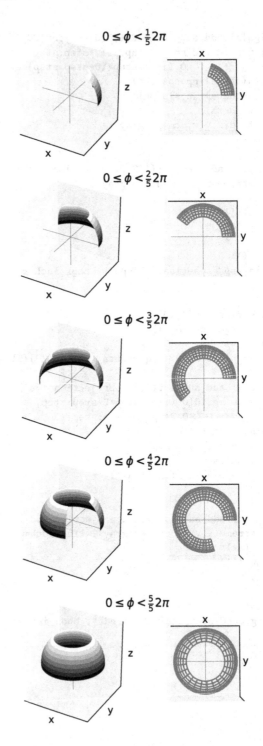

Figure S2

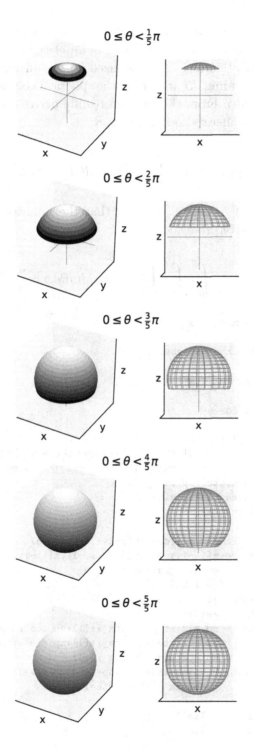

Figure S3

Sometimes, we may deal with a patch of surface area on a sphere. For example, we may calculate the flux of an electric or magnetic field through a spherical surface. As visualized below, an infinitesimal area on a sphere, subtending $d\theta$ and $d\phi$ at the spherical coordinates (r, θ, ϕ), is $da = r^2 \sin\theta d\theta d\phi$. From this, we can readily derive a formula for the surface area of a sphere whose radius is R:

$$A_{\text{sphere}} = \int_0^\pi \int_0^{2\pi} R^2 \sin\theta d\theta d\phi = 4\pi R^2.$$

For the volume of a sphere, we note that the infinitesimal volume is $dV = dadr = r^2 \sin\theta dr d\theta d\phi$, so

$$V_{\text{sphere}} = \int_0^R \int_0^\pi \int_0^{2\pi} r^2 \sin\theta dr d\theta d\phi = \frac{4}{3}\pi R^3.$$

```
# Code Block Appendix C.3

# Visualize a patch of a sphere.

def plot_patch_3D (ax,lim=1):
    r = 1
    step = np.pi/16

    theta_i, phi_j = 5, 0
    theta, phi = np.mgrid[(theta_i-1)*step:(theta_i+3)*step:step,
                          (phi_j-1)*step:(phi_j+3)*step:step]
    x = r*np.sin(theta)*np.cos(phi)
    y = r*np.sin(theta)*np.sin(phi)
    z = r*np.cos(theta)

    ax.plot_wireframe(x,y,z,color='gray',alpha=0.5)
    ax.plot_surface(x[1:-1,1:-1],y[1:-1,1:-1],z[1:-1,1:-1],
                    color='gray',alpha=0.5)
    line = np.array([0,1.05])
    for i in range(2):
        for j in range(2):
            xline = np.sin(step*(theta_i+i))*np.cos(step*(phi_j+j))
            yline = np.sin(step*(theta_i+i))*np.sin(step*(phi_j+j))
            zline = np.cos(step*(theta_i+i))
            ax.plot3D(xline*line, yline*line, zline*line,
                      'black', linewidth=1)

    ax.set_xlim((-lim,lim))
    ax.set_ylim((-lim,lim))
```

```
        ax.set_zlim((-lim,lim))
        ticks = (-lim,0,lim)
        ax.set_xticks(ticks)
        ax.set_yticks(ticks)
        ax.set_zticks(ticks)
        ax.set_xlabel('x')
        ax.set_ylabel('y')
        ax.set_zlabel('z')
        ax.set_box_aspect([1,1,1])

        return

fig = plt.figure(figsize=(8,4))

ax = fig.add_subplot(1,2,1,projection='3d')
plot_patch_3D(ax,lim=1)

ax = fig.add_subplot(1,2,2,projection='3d')
plot_patch_3D(ax,lim=1)
ax.set_ylabel('')
ax.set_yticks(())
ax.view_init(0,-90)

plt.tight_layout()
plt.savefig('fig_appC_spherical_patch.pdf', bbox_inches="tight")
plt.show()
```

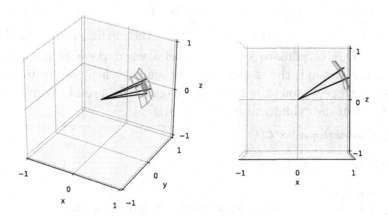

Figure S4

Sometimes, we may deal with a spherically distributed source for an electromagnetic field and calculate the field at a specified position away from the source. The spherical coordinates of a source point are denoted as (r_s, θ_s, ϕ_s), while the observation point is at (r, θ, ϕ). The distance between the source and the observation points is $\sqrt{r^2 + r_s^2 - 2rr_s \left(\sin\theta \sin\theta_s \cos(\phi - \phi_s) + \cos\theta \cos\theta_s \right)}$. In a special case where the observation point lies on the z-axis, corresponding to $\theta = 0$, the above expression for the distance simplifies to $\sqrt{r^2 + r_s^2 - 2rr_s \cos\theta_s}$, which can also be verified with the law of cosine.

Another example of a curved coordinate system in three dimensions is a cylindrical coordinate system, where a point is specified with (r, ϕ, z). r is the distance from the z-axis to the point. Sometimes, another symbol ρ is used instead, in order to distinguish it from the radial distance from the origin in the spherical coordinate system. ϕ is the polar angle measured on the xy-plane just like ϕ in the spherical coordinate. The last coordinate z is the projection of the point onto the z-axis just like z in the Cartesian coordinate system. Hence, the relationship between the Cartesian and cylindrical coordinates is given by:

$$
\begin{aligned}
x &= r \cos\phi \\
y &= r \sin\phi \\
z &= z.
\end{aligned}
$$

The distance between two points (r, ϕ, z) and (r_s, ϕ_s, z_s) is $\sqrt{r^2 + r_s^2 - 2rr_s \cos(\phi - \phi_s) + (z - z_s)^2}$ in the cylindrical coordinate system. This coordinate system is useful when there is a rotational symmetry around the z-axis. For example, when we consider an electrical current along the z-axis, its magnetic field can be conveniently described in the cylindrical coordinate system.

```
# Code Block Appendix C.4

# Cylindrical coordinates
r = 1
step = np.pi/25
z, phi = np.mgrid[-0.9:0.9+step:step, 0:2*np.pi+step:step]
x = r*np.cos(phi)
y = r*np.sin(phi)

fig = plt.figure(figsize=(3,3))
ax = fig.add_subplot(projection='3d')
ax.plot_surface(x,y,z,cmap='gray')
tidy_axis_3d(ax)
```

```
plt.tight_layout()
plt.savefig('fig_appC_cylinder.pdf', bbox_inches="tight")
plt.show()
```

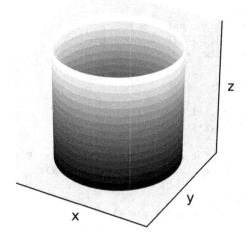

Figure S5

APPENDIX D: FIELD LINES

Throughout this book, we visualized a vector field with a convention of drawing arrows at numerous sample points in the space. We extensively used the `quiver()` function to set each arrow's direction and length. Another visualization is to draw field lines, which are continuous contours whose tangent lines point in the direction of the vector field, and the density of the lines indicates the field strength. As an example, the electric field lines of an electric dipole, two equal and opposite charges with a small separation, are drawn below with the `streamplot()` function. The electric field lines start from the positive charge and end at the negative charge. The field lines are more densely packed where the fields are stronger. You may modify the `charges` variable in the code and visualize the field lines of different point charge distributions.

```
# Code Block Appendix D
import numpy as np
import matplotlib.pyplot as plt
```

```python
# Function to determine electric field
def E(q, q_position, x, y):
    q_x, q_y = q_position[0], q_position[1]
    d3 = np.hypot(x-q_x,y-q_y)**3
    return q*(x-q_x)/d3, q*(y-q_y)/d3

# Grid of x, y points
nx, ny = 64, 64
x = np.linspace(-4, 4, nx)
y = np.linspace(-3, 3, ny)
X, Y = np.meshgrid(x, y)

# Charge locations.
charges = []
charges.append((+1,(-1.5,-0.3)))
charges.append((-1,(+1.5,+0.3)))

# Prepare an array for electric field, E = (Ex, Ey).
Ex, Ey = np.zeros((ny, nx)), np.zeros((ny, nx))

for charge in charges:
    ex, ey = E(*charge,x = X,y = Y)
    Ex = Ex + ex
    Ey = Ey + ey

fig = plt.figure(figsize =(4, 3))

plt.streamplot(x, y, Ex, Ey, color = 'black', linewidth = 1,
               density = 0.25, arrowstyle ='-', arrowsize = 1.5,
               broken_streamlines=False)

# Add filled circles for the charges themselves
charge_colors = {True: '#AAAAAA', False: '#000000'}
for q, pos in charges:
    plt.scatter(pos[0], pos[1], s=200,
                color=charge_colors[q>0], zorder=2)

plt.xlim(-4, 4)
plt.ylim(-3, 3)
plt.axis('equal')
plt.axis('off')
plt.savefig('fig_appD_field_lines.pdf')
plt.show()
```

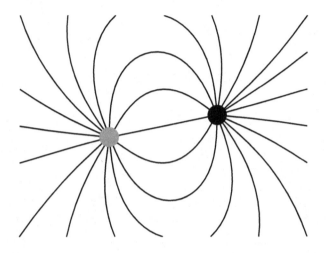

Figure S6

Epilogue

This is just the beginning of a deeper study of electrodynamics. Even though the fundamental laws of electromagnetism can be captured in just a few mathematical relationships (namely, Maxwell's Equations and Lorentz Force Law), the variety of physical phenomena that manifest out of them is stunningly diverse. So diverse that electricity, magnetism, and optics had been considered distinct and separate phenomena. It took many generations of scientists to unify them into a single coherent framework as we know it today. With the advance of our knowledge in electromagnetism, we have seen further development of new technologies (electronics, lasers, etc.), methodologies (microscopy, spectroscopy, etc.), and understanding of the universe (quantum mechanics, field theory, cosmology, etc.). A physicist Richard Feynman noted that "From a long view of the history of mankind – seen from, say, ten thousand years from now – there can be little doubt that the most significant event of the 19th century will be judged as Maxwell's discovery of the laws of electrodynamics."* We enjoyed writing this book, and we hope the readers found our presentation of electrodynamics enjoyable, too.

---

*This quote is from Chapter 1 of Volume 2 of The Feynman Lectures on Physics.

Index

Printed in the United States
by Baker & Taylor Publisher Services